EXTRACTION TECHNIQUES IN ANALYTICAL SCIENCES

Analytical Techniques in the Sciences (AnTS)
Series Editor: David J. Ando, Consultant, Dartford, Kent, UK

A series of open learning/distance learning books which covers all of the major analytical techniques and their application in the most important areas of physical, life and materials sciences.

Titles available in the Series

Analytical Instrumentation: Performance Characteristics and Quality
Graham Currell, University of the West of England, Bristol, UK

Fundamentals of Electroanalytical Chemistry
Paul M.S. Monk, Manchester Metropolitan University, Manchester, UK

Introduction to Environmental Analysis
Roger N. Reeve, University of Sunderland, UK

Polymer Analysis
Barbara H. Stuart, University of Technology, Sydney, Australia

Chemical Sensors and Biosensors
Brian R. Eggins, University of Ulster at Jordanstown, Northern Ireland, UK

Methods for Environmental Trace Analysis
John R. Dean, Northumbria University, Newcastle, UK

Liquid Chromatography–Mass Spectrometry: An Introduction
Robert E. Ardrey, University of Huddersfield, UK

Analysis of Controlled Substances
Michael D. Cole, Anglia Polytechnic University, Cambridge, UK

Infrared Spectroscopy: Fundamentals and Applications
Barbara H. Stuart, University of Technology, Sydney, Australia

Practical Inductively Coupled Plasma Spectroscopy
John R. Dean, Northumbria University, Newcastle, UK

Bioavailability, Bioaccessibility and Mobility of Environmental Contaminants
John R. Dean, Northumbria University, Newcastle, UK

Quality Assurance in Analytical Chemistry
Elizabeth Prichard and Vicki Barwick, LGC, Teddington, UK

Extraction Techniques in Analytical Sciences
John R. Dean, Northumbria University, Newcastle, UK

Forthcoming Titles

Practical Raman Spectroscopy: An Introduction
Peter Vandenabeele, Ghent University, Belgium

Techniques of Modern Organic Mass Spectrometry
Bob Ardrey, Alex Allan and Pete Ashton, Triple A Forensics, Ltd, Oldham, UK

Forensic Analysis Techniques
Barbara H. Stuart, University of Technology, Sydney, Australia

EXTRACTION TECHNIQUES IN ANALYTICAL SCIENCES

John R. Dean

The Graduate School and School of Applied Sciences
Northumbria University, Newcastle, UK

A John Wiley and Sons, Ltd., Publication

Library of Congress Cataloging-in-Publication Data

Record on file

A catalogue record for this book is available from the British Library.

Cloth – 9780470772850   Paper 9780470772843

Set in 10/12pt Times by Laserwords Private Limited, Chennai, India.

To Lynne, Sam and Naomi (and the border terrier, Emmi) for allowing me the time to sit and write this book

Contents

Series Preface

There has been a rapid expansion in the provision of further education in recent years, which has brought with it the need to provide more flexible methods of teaching in order to satisfy the requirements of an increasingly more diverse type of student. In this respect, the *open learning* approach has proved to be a valuable and effective teaching method, in particular for those students who for a variety of reasons cannot pursue full-time traditional courses. As a result, John Wiley & Sons, Ltd first published the *Analytical Chemistry by Open Learning* (ACOL) series of textbooks in the late 1980s. This series, which covers all of the major analytical techniques, rapidly established itself as a valuable teaching resource, providing a convenient and flexible means of studying for those people who, on account of their individual circumstances, were not able to take advantage of more conventional methods of education in this particular subject area.

Following upon the success of the ACOL series, which by its very name is predominately concerned with Analytical *Chemistry*, the *Analytical Techniques in the Sciences* (AnTS) series of open learning texts has been introduced with the aim of providing a broader coverage of the many areas of science in which analytical techniques and methods are now increasingly applied. With this in mind, the AnTS series of texts seeks to provide a range of books which will cover not only the actual techniques themselves, but *also* those scientific disciplines which have a necessary requirement for analytical characterization methods.

Analytical instrumentation continues to increase in sophistication, and as a consequence, the range of materials that can now be almost routinely analysed has increased accordingly. Books in this series which are concerned with the *techniques* themselves will reflect such advances in analytical instrumentation, while at the same time providing full and detailed discussions of the fundamental concepts and theories of the particular analytical method being considered. Such books will cover a variety of techniques, including general instrumental analysis, spectroscopy, chromatography, electrophoresis, tandem techniques, electroanalytical methods, X-ray analysis and other significant topics. In addition, books in the series will include the *application* of analytical techniques in areas such as environmental science, the life sciences, clinical analysis, food science, forensic analysis, pharmaceutical science, conservation and archaeology, polymer science and general solid-state materials science.

Written by experts in their own particular fields, the books are presented in an easy-to-read, user-friendly style, with each chapter including both learning objectives and summaries of the subject matter being covered. The progress of the reader can be assessed by the use of frequent self-assessment questions (SAQs) and discussion questions (DQs), along with their corresponding reinforcing or remedial responses, which appear regularly throughout the texts. The books are thus eminently suitable both for self-study applications and for forming the basis of industrial company in-house training schemes. Each text also contains a large amount of supplementary material, including bibliographies, lists of acronyms and abbreviations, and tables of SI Units and important physical constants, plus where appropriate, glossaries and references to literature sources.

It is therefore hoped that this present series of textbooks will prove to be a useful and valuable source of teaching material, both for individual students and for teachers of science courses.

Dave Ando
Dartford, UK

Preface

This book introduces a range of extraction techniques as applied to the recovery of organic compounds from a variety of matrices. In line with other texts in the Analytical Techniques in the Sciences (AnTS) Series, discussion and self-assessment questions provide the reader with the opportunity to assess their own understanding of aspects of the text. This book has been designed to be 'user-friendly' with illustrations to aid understanding. This text is arranged into thirteen chapters as follows.

Chapter 1 introduces all the key aspects that need to be considered, pre- and post-extraction. In particular, it highlights the range of organic compounds that are extracted in analytical sciences. This chapter then addresses pre-sampling issues by way of a desk-top study of a contaminated land site using historic maps. Specific sampling strategies for solid, aqueous and air samples are considered. The natural progression in any analytical protocol would then be to carry out the extraction technique. However, as the rest of the book details how to perform different extractions no details are provided at this point. Post-extraction details focus on the main chromatographic approaches for analysing organic compounds, i.e. gas chromatography and high performance liquid chromatography. Both techniques are covered from a practical perspective. Issues around sample pre-concentration post-extraction are also discussed in terms of the most popular approaches used. Finally, quality assurance aspects and health and safety issues are considered.

Chapter 2 considers the classical approaches for extracting organic compounds from aqueous samples, namely liquid–liquid extraction (LLE). Details of the basic theory applicable to LLE are explained together with important practical aspects, including choice of solvents, the apparatus and procedure to undertake LLE and practical problems and remedies for undertaking LLE. Finally, the specific extraction technique of purge and trap and its application for recovering volatile organic compounds from aqueous samples is explained.

Chapter 3 considers the use of solid phase extraction (or SPE) for the recovery of organic compounds from aqueous samples. The different types of SPE media are considered as well as the different formats in which SPE can be performed, solvent selection and factors influencing SPE. The five main aspects of SPE operation are reviewed both generically and then via a series of applications using normal phase, reversed phase, ion exchange and molecularly imprinted polymers. Finally, the use of automated and in-line SPE is considered using a selected example.

Chapter 4 considers the use of solid phase microextraction (or SPME) for the recovery of organic compounds from aqueous samples (although mention is also made of its applicability for headspace sampling), followed by either GC or HPLC. The practical aspects of using the fibres are described in detail as well as their applicability for a range of sample types in different modes of operation.

Chapter 5 describes new developments in microextraction. Particular developments highlighted include stir-bar sorptive extraction (SBSE), liquid phase microextraction (specifically, single drop microextraction (SDME)), membrane microextraction (specifically, the semipermeable membrane device (SPMD), the polar organic chemical integrative sampler (POCIS), 'Chemcatcher', the ceramic dosimeter and membrane enclosed-sorptive coating (MESCO)), as well as microextraction in a packed syringe (MEPS).

Chapter 6 considers the classical approaches for extracting organic compounds from solid samples, namely Soxhlet extraction (LLE). Practical guidance on the use of Soxhlet extraction is provided along with choice of solvent, and the apparatus and procedure to undertake extraction. In addition, automated Soxhlet (or 'Soxtec') extraction is discussed alongside other approaches that utilize sonication or shake-flask extraction for the recovery of organic compounds from solid matrices.

Chapter 7 describes the use of pressurized fluid extraction (PFE) (also known as accelerated solvent extraction or pressurized liquid extraction) for the recovery of organic compounds from solid matrices. The theoretical aspects of the approach are described, as well as the range of commercial apparatus that is currently available. Approaches for method development for PFE are described, as well as a range of applications including approaches for parameter optimization, *in situ* clean-up (also known as selective PFE) and shape selective, fractionation PFE.

Chapter 8 describes the use of microwave-assisted extraction (MAE) for the recovery of organic compounds from solid matrices. Instrumentation for both atmospheric and pressurized MAE are highlighted, with the latter dominating in its applicability. A range of applications is considered, as well as some recommendations on the use of MAE in analytical sciences.

Chapter 9 considers developments in matrix solid phase dispersion (MSPD) for solid samples. The procedure for performing MSPD is highlighted, as well as its applicability to a range of sample types. A range of factors that can influence

MSPD is then discussed. Finally, a comparison between MSPD and solid phase extraction is made.

Chapter 10 describes the technique of supercritical fluid extraction (SFE). After an initial description of what is a supercritical fluid, the option of carbon dioxide as the fluid of choice is discussed. A detailed description of the instrumentation for SFE is outlined, together with the options for adding modifiers to the system. Finally, a range of applications for SFE in analytical sciences is described.

Chapter 11 considers the analysis of volatile organic compounds (VOCs) in gaseous samples. A discussion on the techniques for air sampling, including whole air collection in containers, enrichment into solid sorbents (active and passive sampling), desorption techniques and on-line sampling, is also included.

Chapter 12 includes a detailed discussion on the important extraction method criteria, namely, sample mass/volume, extraction time, solvent type and consumption, extraction method, sequential or simultaneous extraction, method development time, operator skill, equipment cost, level of automation and extraction method approval. This chapter then considers the above criteria in the context of comparing extraction techniques for (semi-) solid samples and liquid samples. A comparison is also made of the approaches for air samples. In addition, this chapter also considers the role and use of certified reference materials.

The final chapter (Chapter 13) considers the resources available when considering the use of extraction techniques in analytical sciences. The role of the Worldwide Web in accessing key sources of information (publishers, companies supplying instrumentation and consumables, institutions and databases) is highlighted.

John R. Dean
Northumbria University, Newcastle, UK

Acknowledgements

This present text includes material which has previously appeared in three of the author's earlier books, i.e. *Extraction Methods for Environmental Analysis* (1998), *Methods for Environmental Trace Analysis* (AnTS Series, 2003) and *Bioavailability, Bioaccessibility and Mobility of Environmental Contaminants* (AnTS Series, 2007), all published by John Wiley & Sons, Ltd. The author is grateful to the copyright holders for granting permission to reproduce figures and tables from his three earlier publications.

Dr Marisa Intawongse is acknowledged for her assistance with the compilation of Chapters 3 and 4. Dr Pinpong Kongchan is thanked for the drawing of Figures 6.3, 8.2, 8.3, 8.5 and 8.6, Dr Michael Deary for providing Figure 1.1 and Naomi Dean for the drawing of Figures 1.5 and 1.6.

The front cover shows a photograph of Sycamore Gap located on Hadrian's Wall in Northumberland, UK, where the tree, sky and ground symbolize the areas of soil, air and water aspects of this book. This location was used in the 1991 film 'Robin Hood Prince of Thieves' starring Kevin Costner and so to my family it is known as 'Robin's tree' – Robin Hood is also immortalized in my family with the phrase 'after them you hools!'. Picture provided by John R. Dean, Northumbria University, Newcastle, UK.

Acronyms, Abbreviations and Symbols

ACN	acetonitrile
ACS	American Chemical Society
AOAC	Association of Official Analytical Chemists
APCI	atmospheric pressure chemical ionization
ASE	accelerated solvent extraction
ASTM	American Society for Testing and Materials
BAM	The Federal Institute for Materials Research and Testing
BCR	Community Bureau of Reference
BNAs	bases, neutral species, acids
BTEX	benzene, toluene, ethylbenzene and xylenes
CAR	carboxen
CI	chemical ionization
COSHH	Control of Substances Hazardous to Health
CRM	certified reference material
DCM	dichloromethane
DIN	Deutsches Institut für Normung
DVB	divinylbenzene
ECD	electron capture detector
EI	electron impact
ES	electrospray
EU	European Union
EVACS	evaporative concentration system
FDA	Food and Drug Administration
FID	flame ionization detector
GC	gas chromatography

HPLC	high performance liquid chromatography
HS	headspace
HTML	hypertext markup language
ICP	inductively coupled plasma
ID–GC–MS	isotope dilution–gas chromatography–mass spectrometry
IR	infrared
IRMM	Institute for Reference Materials and Measurements
IT–MS	ion trap–mass spectrometry
LC	liquid chromatography
LDPE	low-density polyethylene
LGC	Laboratory of the Government Chemist
LLE	liquid–liquid extraction
LOD	limit of detection
LOQ	limit of quantitation
MAE	microwave accelerated extraction
MCL	maximum concentration level
MEPS	microextraction in a packed syringe
MESCO	membrane enclosed-sorptive coating
MIP	molecularly imprinted polymer
MS	mass spectrometry
MSD	mass selective detector
MSPD	matrix solid phase dispersion
NIST	National Institute of Science and Technology
NMIJ	National Metrology Institute of Japan
NP (HPLC)	normal phase (high performance liquid chromatography)
NRC	National Research Council (of Canada)
NRCCRM	National Research Centre for Certified Reference Materials
NWRI	National Water Research Institute
ODS	octadecylsilane
PAHs	polycyclic aromatic hydrocarbons
PCBs	polychlorinated biphenyls
pdf	portable document format
PDMS	polydimethylsiloxane
PEEK	poly(ether ether ketone)
PFAs	perfluoroalkoxy fluorocarbons
PFE	pressurized fluid extraction
PHWE	pressurized hot water extraction
PLE	pressurized liquid extraction
POCIS	polar organic chemical integrative sampler
POPs	persistent organic pollutants
ppb	parts per billion (10^9)

ppm	parts per million (10^6)
ppt	parts per thousand (10^3)
PSE	pressurized solvent extraction
PTV	programmed temperature vaporizer
PVC	poly(vinyl chloride)
QA	quality assurance
RAM	restricted access media
RP (HPLC)	reversed phase (high performance liquid chromatography)
RSC	The Royal Society of Chemistry
RSD	relative standard deviation
SCX	strong cation exchange
SBSE	stir-bar sorptive extraction
SDME	single drop microextraction
SFC	supercritical fluid chromatography
SFE	supercritical fluid extraction
SIM	single (or selected) ion monitoring
SPE	solid phase extraction
SPLE	selective pressurized liquid extraction
SPMD	semipermeable membrane device
SPME	solid phase microextraction
SSSI	site of special scientific interest
SI (units)	Système International (d'Unitès) (International System of Units)
TFM	tetrafluoromethoxy (polymer)
TIC	total ion current
TOF–MS	time-of-flight–mass spectrometry
TSD	thermionic specific detector
URL	uniform resource locator
USEPA	United States Environmental Protection Agency
UV	ultraviolet
VOCs	volatile organic compounds
WWW	Worldwide Web

c	speed of light; concentration
D	distribution ratio
E	energy; electric field strength
f	(linear) frequency
I	electric current
K_d	distribution coefficient
K_{ow}	octanol–water partition coefficient
$\log P$	log of octanol–water partition coefficient
m	mass

P	pressure
R	molar gas constant
t	time; Student factor
V	electric potential
z	ionic charge
λ	wavelength
ν	frequency (of radiation)
σ	measure of standard deviation
σ^2	variance

About the Author

John R. Dean, B.Sc., M.Sc., Ph.D., D.I.C., D.Sc., FRSC, C.Chem., C.Sci., Cert. Ed., Registered Analytical Chemist

John R. Dean took his first degree in Chemistry at the University of Manchester Institute of Science and Technology (UMIST), followed by an M.Sc. in Analytical Chemistry and Instrumentation at Loughborough University of Technology, and finally a Ph.D. and D.I.C. in Physical Chemistry at the Imperial College of Science and Technology (University of London). He then spent two years as a postdoctoral research fellow at the Food Science Laboratory of the Ministry of Agriculture, Fisheries and Food in Norwich, in conjunction with the Polytechnic of the South West in Plymouth (now the University of Plymouth). His work there was focused on the development of directly coupled high performance liquid chromatography and inductively coupled plasma–mass spectrometry methods for trace element speciation in foodstuffs. This was followed by a temporary lectureship in Inorganic Chemistry at Huddersfield Polytechnic (now the University of Huddersfield). In 1988, he was appointed to a lectureship in Inorganic/Analytical Chemistry at Newcastle Polytechnic (now Northumbria University). This was followed by promotion to Senior Lecturer (1990), Reader (1994), Principal Lecturer (1998) and Associate Dean (Research) (2004). He was also awarded a personal chair in 2004. In 2008 he became the Director of The Graduate School at Northumbria University as well as Professor of Analytical and Environmental Sciences in the School of Applied Sciences.

In 1998, he was awarded a D.Sc. (University of London) in Analytical and Environmental Science and was the recipient of the 23rd Society for Analytical Chemistry (SAC) Silver Medal in 1995. He has published extensively in analytical and environmental science. He is an active member of The Royal Society of Chemistry (RSC) Analytical Division, having served as a member of the Atomic

Spectroscopy Group for 15 years (10 as Honorary Secretary) as well as a Past Chairman (1997–1999). He has served on the RSC Analytical Division Council for three terms and is a former Vice-President (2002–2004), as well as a past-Chairman of the North-East Region of the RSC (2001–2003).

Chapter 1

Pre- and Post-Extraction Considerations

Learning Objectives

- To appreciate the wide ranging types of organic compounds that are investigated in environmental and food matrices.
- Using an example, to be aware of pre-sampling issues associated with a contaminated land site.
- To be aware of the information required for a desk-top study (in a contaminated land situation).
- To understand the different sampling strategies associated with solid, aqueous and air samples.
- To be aware of the different types of contaminant distribution on a site.
- To understand the practical aspects of soil and sediment sampling.
- To understand the practical aspects of water sampling.
- To understand the practical aspects of air sampling.
- To be aware of the different analytical techniques available to analyse organic compounds.
- To understand and explain the principle of operation of a gas chromatography system.
- To understand and explain the principle of operation of a high performance liquid chromatography system.
- To be able to understand the principles of quantitative chromatographic analysis.

Extraction Techniques in Analytical Sciences   John R. Dean
© 2009 John Wiley & Sons, Ltd

- To be aware of the approaches and limitations for sample pre-concentration in the analysis of organic compounds.
- To appreciate the importance of quality assurance in quantitative analysis.
- To understand the health and safety aspects of performing laboratory work and the consequences for non-compliance.

1.1 Introduction

This book is concerned with the removal of organic compounds, principally persistent organic compounds (POPs), from a range of sample matrices including environmental matrices (soil, water and air samples), but also some other matrices including foodstuffs. The book is designed to be an informative guide to a range of extraction techniques that are used to remove organic compounds from various matrices. The use of discussion questions (DQs) and self-assessment questions (SAQs) throughout the text should allow you (the reader) to think about the main issues and to allow you to consider alternative approaches.

1.2 Organic Compounds of Interest

The range of organic compounds of interest in the environment and in other matrices varies enormously. They range from simple aromatic cyclic structures, for example, benzene, toluene, ethylbenzene and xylene(s) (collectively known as BTEX), to larger molecular weight compounds, such as polycyclic aromatic hydrocarbons (PAHs), and more complicated structures, e.g. pesticides and polychlorinated biphenyls (PCBs). A list of organic compounds that are measured in environmental (and other) matrices is shown in Table 1.1.

SAQ 1.1

What are the important physical and chemical properties of these organic compounds that are useful to know when extracting them from sample matrices?

1.3 Pre-Sampling Issues

Prior to sampling it is necessary to consider a whole range of issues that are directly/indirectly going to influence the quality of the final data that is produced after what is often a long and costly process. Therefore it is imperative to think

Table 1.1 Potential organic contaminants in the environment

Class of compound	Name of specific compound
Aromatic hydrocarbons	Benzene
	Chlorophenols
	Ethylbenzene
	Phenol
	Toluene
	o-xylene
	m,*p*-xylene
	Polycyclic aromatic hydrocarbons
Chlorinated aliphatic hydrocarbons	Chloroform
	Carbon tetrachloride
	Vinyl chloride
	1,2-Dichloroethane
	1,1,1-Trichloroethane
	Trichloroethene
	Tetrachloroethene
	Hexachlorobuta-1,3-diene
	Hexachlorocyclohexanes
	Dieldrin
Chlorinated aromatic hydrocarbons	Chlorobenzenes
	Chlorotoluenes
	Pentachlorophenol
	Polychlorinated biphenyls
	Dioxins and furans

about the 'whole picture' before any sampling is started. In reality a range of individuals will be involved in the process. To illustrate some of the steps involved a simple generic approach is presented to allow you to think about the overall process.

DQ 1.1

Is a former industrial site suitable for building domestic houses?

Answer

[In order to answer this it is appropriate to consider yourself as the individual responsible for overseeing this work on behalf of the current owner of the land.]

Initial thoughts should revolve around carrying out a desk-top study. A desk-top study, as the name suggests, involves gathering information that is readily available without necessarily having to analyse anything (at least at this point in time). A desk-top study may contain the following information:

- Physical setting.
  - Site details including a description of location, map reference, access to site, current land use and general description of site.
- Environmental setting.
  - Site geology including a description of surface and below-surface geology, e.g. coal seam.
  - Site hydrogeology including details of river or stream flows and whether groundwater is abstracted and for what purpose.
  - Site hydrology including known rainfall and river/stream/pond locations.
  - Site ecology and archaeology including whether the site has any known scheduling, e.g. site of special scientific interest (SSSI); any features of archaeological significance.
  - Mining assessment, e.g. evidence of former quarrying activity.
- Industrial setting and recent site history. Information available via historic and modern ordnance survey maps including (aerial) photographs of the site.
- Qualitative risk assessment including development of a site-specific conceptual model that seeks to assess the following:
  - Source of contaminants.
  - The pathway by which a contaminant could come into contact with a receptor, e.g. people.
  - The characteristics and sensitivity of the receptor to the contaminant.
- Site walkover, i.e. by visiting the site it is possible to identify key issues, major features, position of walkways, etc.
- Any previous site investigations.
- Conclusions and recommendations.

Useful information can be gathered about a former industrial site by obtaining detailed historic ordnance survey maps. By studying these maps it will be evident what building infrastructure will have been present at set times in history. For example, Figure 1.1(a) shows a historic map (1898) from a site which is largely marsh land and was underdeveloped in 1898, while Figures 1.1(b–d) illustrate the growth of the industrial aspects of the site from 1925 (Figure 1.1(b)) through

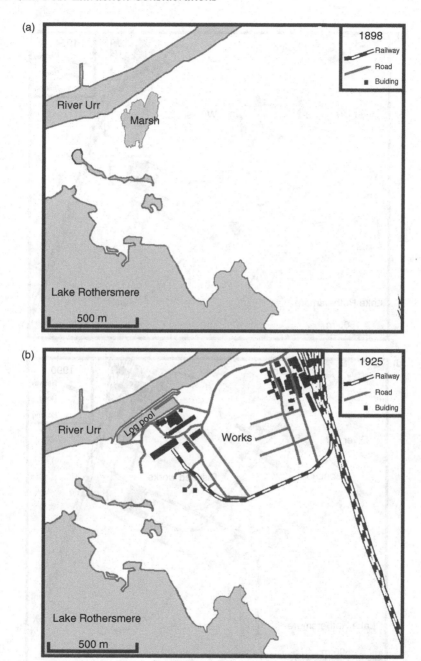

Figure 1.1 Historic maps of a selected site: (a) 1898; (b) 1925; (c) 1954; (d) 1990. Reproduced by permission of Dr M. Deary, Northumbria University, Newcastle, UK.

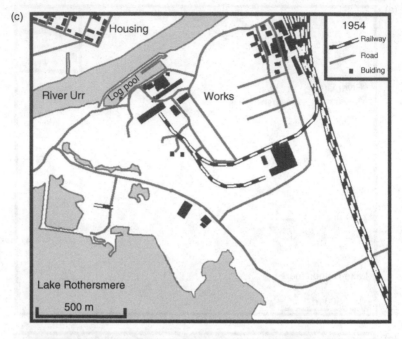

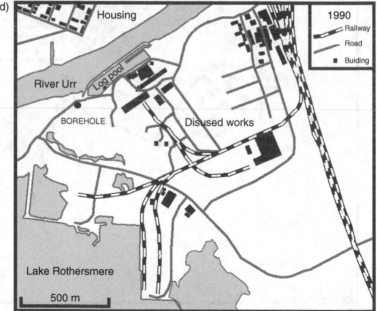

Figure 1.1 (*continued*)

to 1954 (Figure 1.1(c)) and its subsequent decline by 1990 (Figure 1.1(d)). The emergent development of housing is noted in Figure 1.1(d). In addition, information about the use of the former buildings can be obtained from local archivists, e.g. city/town councils and history societies, who will retain records on historic activities. By gathering this detailed information it is possible to build up a picture of possible organic contaminants that may still be present on the site (not necessarily amenable on the surface but buried beneath other material).

DQ 1.2

What other contaminants may be present on the site?

Answer

As well as organic compounds other contaminants may be present, including heavy metals, asbestos etc.

With regard to carrying out some specific sampling it is necessary to obtain answers, in advance, about the following:

(1) Do you have permission to obtain samples from the site?

(2) Is specialized sampling equipment required? If so, do you have access to it? If not can you obtain the equipment and from whom?

(3) How many samples (including replicates) will it be necessary to take?

(4) What soil/water/air testing is required?

(5) What instrumentation is available to do the testing on?

(6) Is the instrumentation limited with respect to sample size (mass or volume)? Does sample size constrain the analytical measurement?

(7) What quality assurance procedures are available? Has a protocol been developed?

(8) What types of container are required to store the samples and do you have enough of them?

(9) Do the containers require any pre-treatment/cleaning prior to use and will this be done in time?

(10) Is any sample preservation required? If so what is it and how might it impact on the analysis of the contaminants?

1.4 Sampling Strategies: Solid, Aqueous and Air Samples

Ideally, all sample matrices should be analysed at or on-site without any need to transport samples to a laboratory. Unfortunately in most cases this does not happen and samples are transported back to a laboratory and analysed. The exception is where a preliminary assessment takes place on site, for example, by using a photoionization detector to assess the level of volatile organic compounds in the atmosphere. The issue in most instances is to consider how many samples should be taken and from which location. Therefore significant consideration needs to be given to the sampling protocol as to whether the sample is solid, liquid or gaseous in order that the data that are obtained at the end of the analytical process has meaning and can be interpreted appropriately. Two main types of sampling can be undertaken: random or purposeful sampling. The former is the most important as it infers no selectivity in the sampling process.

The sampling process involves the following:

- selection of the sample points;
- the size of the sample area;
- the shape of the sample area;
- the number of sampling units in each sample.

It is advantageous before sampling to consider information, e.g. location of former buildings on the site, to potentially assess the likely distribution of the contaminants. Any distribution of contaminants can be generally described as:

- random;
- uniform (homogenous);
- patchy or stratified (homogenous within sub-areas);
- present as a gradient.

Examples of these potential likely distributions of contaminants are shown in Figure 1.2.

In practice, however, the site to be sampled can be hindered by the occurrence of modern building, footpaths and other infrastructure obstacles (e.g. stanchions for bridges).

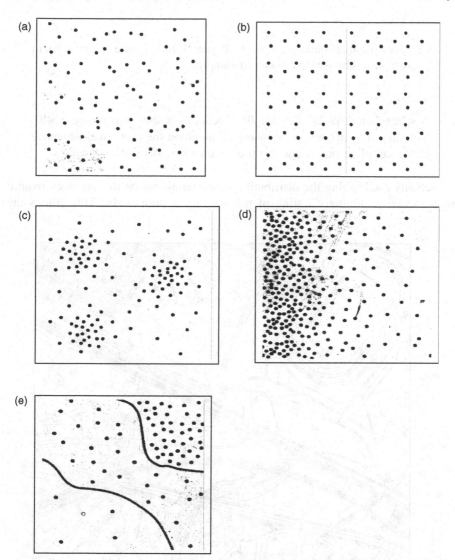

Figure 1.2 Different distributions of inorganic and organic contaminants: (a) random; (b) uniform (homogeneous); (c) patchy; (d) stratified (homogeneous within sub-areas); (e) gradient. From Dean, J. R., *Methods for Environmental Trace Analysis*, AnTS Series. Copyright 2003. © John Wiley & Sons, Limited. Reproduced with permission.

DQ 1.3

Consider the map outline shown in Figure 1.3(a). Based on the current
site where might you to choose to sample?

Answer

A suggestion of particular sampling locations is shown in Figure 1.3(b).
Note that it is not always possible to maintain the numerical sequence
of the sampling points due the presence of permanent structures.

Actually establishing the distribution of contaminants on the site does require
some actual preliminary testing of the site, i.e. a pilot study. This allows the

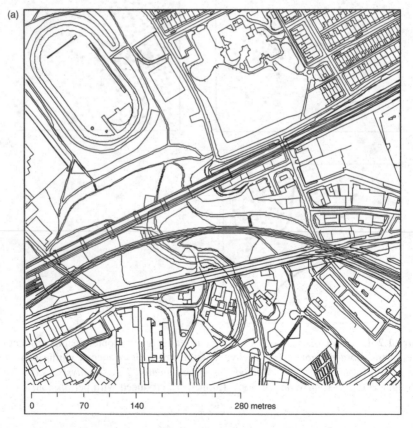

Figure 1.3 An example of a potential contaminated land site for investigation. (a) Consider the options for locating a sample grid. (b) Sampling grid and selected sites (num-
bered). © Crown Copyright Ordnance Survey. An EDIMA Digimap/JISC supplied service.

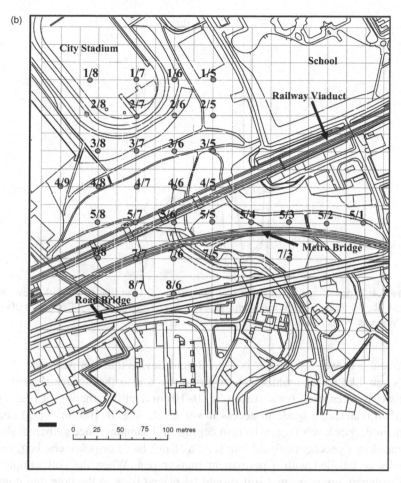

Figure 1.3 (*continued*)

level and distribution of contaminants to be assessed. The sampling position can be assessed by overlaying a 2-dimensional coordinate grid on the site to be investigated (see for example, Figure 1.3(b), and then deciding to sample, for example, from either every grid location or every other grid location. This approach to sampling is appropriate in the context of contaminants which are likely to be homogeneously distributed about the site.

1.4.1 Practical Aspects of Sampling Soil and Sediment

This sample type is often characterized by its heterogeneity and hence diversity of chemical and physical properties. Samples are usually taken with an auger,

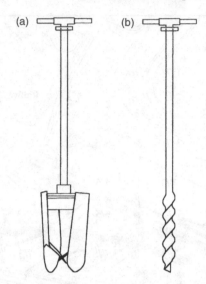

Figure 1.4 Types of augers used for soil sampling: (a) twin blade; (b) corkscrew. From Dean, J. R., *Methods for Environmental Trace Analysis*, AnTS Series. Copyright 2003. © John Wiley & Sons, Limited. Reproduced with permission.

spade and a trowel. The auger is a hand-held device that can penetrate the soil in a screw-like manner which acts to bring the soil to the surface (Figure 1.4). A trowel is often used for surface (e.g. 0–10 cm depth) gathering of previously disturbed material, a spade to access lower levels (e.g. 0–100 cm depth) and an auger for deeper levels (e.g. >100 cm depth). Soil samples, once gathered, should be placed in a geochemical soil bag (e.g. a 'Kraft bag') or polythene bag, sealed and clearly labelled with a permanent marker pen. When the soil sample has been gathered any unwanted soil should be placed back in the hole and covered with a grass sod, if appropriate. The samples are then transported back to the laboratory and dried. In the case of the geochemical soil bag the sample can be left *in-situ* and dried. Drying is normally done by placing the sample in a special drying cabinet that allows air flow at a temperature <30°C.

DQ 1.4

Why should a higher temperature *not* be used for organic compounds?

Answer

Higher temperatures should not be used for samples containing organic compounds to prevent premature loss of the compounds under investigation.

Depending on the sample moisture content the drying process may be complete with 48 h. The air dried sample is then sieved (2 mm diameter holes) through a pre-cleaned plastic sieve to remove stones, large roots and any other unwanted material. The sieved sample can then be sub-sampled and analysed. Sometimes it is appropriate to reduce the sample size further. For example, samples may be sieved through a pre-cleaned 250 µm sieve such that two size fractions are available for analysis, i.e. the >250 µm and <250 µm fractions. The prepared soil samples can then be further sub-sampled using the process of coning and quartering to obtain a representative sample for extraction and subsequent analysis.

SAQ 1.2

What is coning and quartering?

1.4.2 Practical Aspects of Sampling Water

Water can be classified into many types, e.g. surface waters (rivers, lakes, runoff, etc.), groundwaters and springwaters, wastewaters (mine drainage, landfill leachate, industrial effluent, etc.), saline waters, estuarine waters and brines, waters resulting from atmospheric precipitation and condensation (rain, snow, fog, dew), process waters, potable (drinking) waters, glacial melt waters, steam, water for sub-surface injections, and water discharges including waterborne materials and water-formed deposits.

Water is often an heterogeneous substance with both spatial and temporal variation.

DQ 1.5

Why might spatial variation occur in natural water?

Answer

Spatial variation occurs due to stratification within lakes due to variations in flow, chemical composition and temperature.

DQ 1.6

Why might temporal variation occur in natural water?

Answer

Temporal variation, i.e. variation with respect to time occurs, for example, because of heavy precipitation (i.e. snow, rain) and seasonal changes.

A schematic of a typical manual water sampling device is shown in Figure 1.5. The device consists of an open tube with a known volume (e.g. 1 to 30 l) fitted with a closure mechanism at either end. The device is usually made of stainless-steel or PVC. The sample is taken by lowering the device to a pre-determined depth and then opening both ends for a short time. Then, both ends are closed and sealed. By this process the water is sampled at a specified depth. The sampled water is then brought to the surface and transferred to a suitable glass container with a sealable lid.

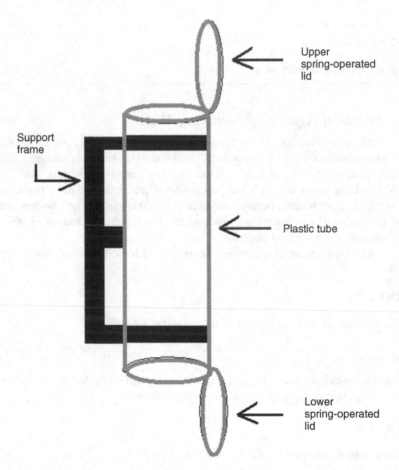

Figure 1.5 A schematic of a typical manual device used for water sampling. Figure drawn and provided by courtesy of Naomi Dean.

> **SAQ 1.3**
>
> Why is it often not advisable to use a plastic container for organic compounds?

Fortunately the methods of preservation are few for organic compounds and intended to fulfil the following criteria: to retard biological action, to retard hydrolysis of chemical compounds and complexes, to reduce volatility of constituents and to reduce adsorption effects. For organic compounds the normal process is to store the water samples for the shortest possible time, in the dark and at 4°C. Suggested storage conditions for selected organic compounds are shown in Table 1.2.

1.4.3 Practical Aspects of Air Sampling

Air sampling can be classified into two distinct themes: vapour/gas sampling or particulate sampling. In the case of the latter, particles are collected on filters (e.g. fibreglass, cellulose fibres) which act as physical barriers whereas in the former case air-borne compounds are trapped on a sorbent (e.g. ion-exchange resins, polymeric substrates) which provide active sites for chemical/physical retention of material.

In sorbent tube sampling (Figure 1.6), volatile and semi-volatile organic compounds are pumped from the air and trapped on the surface of the sorbent (Figure 1.6 (a)). Quantitative sampling is possible by allowing a measured quantity of air (typical volumes of $10–500\,m^3$) to pass through the sorbent. The sorbent tube is then sealed and transported back to the laboratory for analysis. As the organic compounds collected are either volatile or semi-volatile they will be analysed by gas chromatography (see Section 1.5.1). First however, they need to be desorbed by either the use of organic solvent (solvent extraction) or heat (thermal desorption). The latter approach can be done in a fully automated manner using commercial instrumentation and is therefore the preferred analytical approach.

1.5 An Introduction to Practical Chromatographic Analysis

Organic compounds can be analysed by a variety of analytical techniques including chromatographic and spectroscopic methods. However, in this book the main emphasis is on the use of chromatographic approaches. A brief overview of some of the most important chromatographic techniques is provided together with some practical information.

Table 1.2 Selected examples of preservation techniques for water samples[a]

Compound	Storage container	Preservation	Maximum holding time
Pesticides (organochlorine)	Glass	1 ml of a 10 mg ml^{-1} HgCl$_2$ or adding of extraction solvent (500 ml of water)	7 days, 40 days after extraction
Pesticides (organophosphorus)	Glass	1 ml of a 10 mg ml^{-1} HgCl$_2$ or adding of extraction solvent (500 ml of water)	14 days, 28 days after extraction
Pesticides (chlorinated herbicides)	Glass	Cool to 4°C, seal, add HCl to pH < 2 (500 ml of water)	14 days
Pesticides (polar)	Glass	1 ml of a 10 mg ml^{-1} HgCl$_2$ (500 ml of water)	28 days
Phenolic compounds	Glass	Cool to 4°C, add H$_2$SO$_4$ to pH < 2 (500 ml of water)	28 days

[a] As recommended by different agencies (USEPA and ISO).

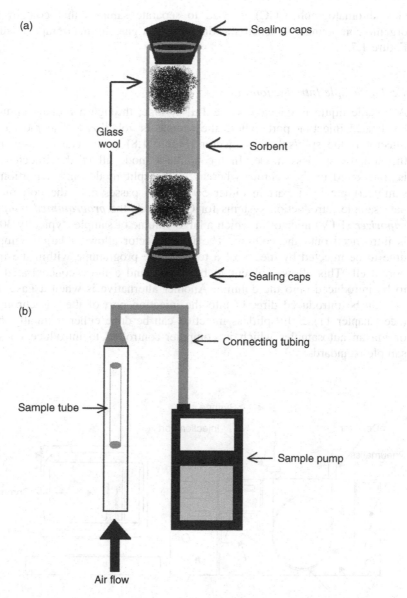

Figure 1.6 Air sampling: (a) schematic of a typical sorbent tube; (b) schematic of the system used to carry out measurements. Figure drawn and provided by courtesy of Naomi Dean.

1.5.1 Gas Chromatography

Gas chromatography (GC) is used to separate samples that contain volatile organic compounds. A schematic diagram of a gas chromatograph is shown in Figure 1.7.

1.5.1.1 Sample Introduction in GC

A volatile liquid is injected, via a 1 µl syringe, through a rubber septum in to the heated injection port, where the sample is volatilized. The most common injector is the split/splitless injector (Figure 1.8) which can operate in either the split or splitless mode. In the splitless mode all of the injected sample is transferred to the column whereas in the split mode only a portion of the sample (typically 1 part in either 50 or 100) passes onto the column. Alternate sample introduction systems for GC include the **programmed temperature vaporizer** (PTV) injector in which a large volume of sample (typically 30–50 µl) is introduced onto the column. The PTV injector allows a larger sample volume to be injected by means of a temperature programme within the injection port itself. This allows solvent to be vented and a more concentrated sample to be introduced onto the column. Another alternative is when a gaseous sample can be introduced directly into the injection port of the gas chromatoraph (see Chapter 11). Split/splitless injection can be done either manually, by hand or via an autosampler which is computer-controlled to introduce consecutive samples/standards.

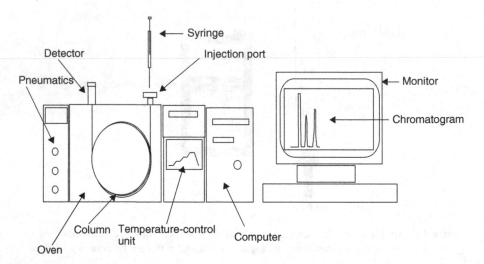

Figure 1.7 Schematic diagram of a typical gas chromatograph. Reproduced by permission of Mr E. Ludkin, Northumbria University, Newcastle, UK.

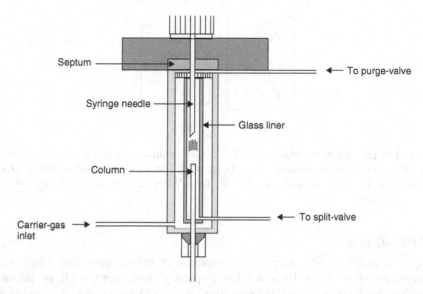

Figure 1.8 Schematic diagram of a split/splitless injector used in gas chromatography. Reproduced by permission of Mr E. Ludkin, Northumbria University, Newcastle, UK.

DQ 1.7

How might you manually inject a sample/standard into the gas chromatograph?

Answer

In the manual injection mode the sample/standard is introduced as follows:

- The syringe is filled (1.0 µl) with the sample/standard solution; this is achieved by inserting the needle of the syringe into the solution and slowly raising and then rapidly depressing the plunger. After several repeats of this process the plunger is raised to the 1.0 µl position on the calibrated syringe.

- The outside of the syringe is then wiped clean with a tissue.

- Then, the syringe is placed into the injector of the gas chromatograph and the plunger on the syringe is rapidly depressed to inject the sample.

A gaseous carrier gas (nitrogen or helium) transports the sample from the injection port to the column.

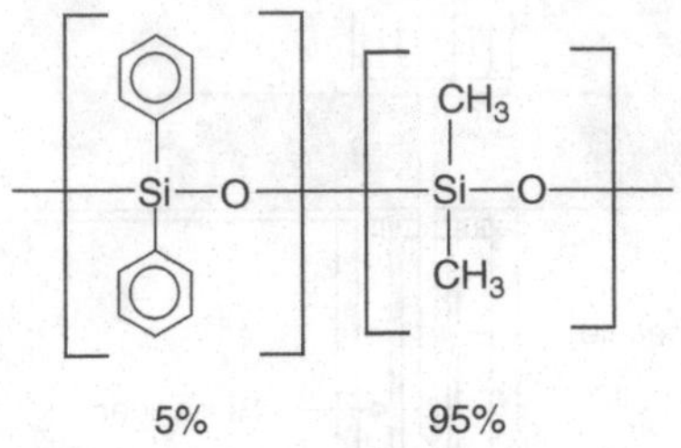

Figure 1.9 The stationary phase of a DB-5 GC column, consisting of 5% diphenyl- and 95% dimethylpolysiloxanes. From Dean, J. R., *Methods for Environmental Trace Analysis*, AnTS Series. Copyright 2003. © John Wiley & Sons, Limited. Reproduced with permission.

1.5.1.2 GC Column

A typical capillary GC column is composed of polyimide-coated silica with dimensions of between 10 and 60 m (typically 30 m) long with an internal diameter between 0.1 and 0.5 mm (typically 0.25 mm), and a crosslinked silicone polymer stationary phase (for example, 5% polydiphenyl–95% polydimethylsiloxane – generically known as a DB-5 column), coated as a thin film on the inner wall of the fused silica (SiO_2) capillary of thickness 0.1–0.5 μm (typically 0.25 μm) (Figure 1.9).

The column is located within an oven, capable of accurate and rapid temperature changes, allowing either isothermal or temperature programmed operation for the separation of organic compounds. In the isothermal mode the temperature of the oven, and hence the column environment, is maintained at a fixed temperature (e.g. typically in the range 70–120°C), while in the temperature programmed mode a more complex heating programme is used. This approach is often necessary for the separation of complex mixtures of organic compounds. A typical oven temperature programme could be as follows: start at an initial temperature of 70°C for 2 min, then a temperature rise of 10°C/min up to 220°C, followed by a 'hold time' of 2 min. In order for the next sample to be introduced the oven must cool back to 70°C prior to injection; this process is rapid, taking approximately 1–2 min.

1.5.1.3 Detection in GC

After GC separation the eluting compounds need to be detected. The most common detectors for GC are the universal detectors, as follows:

- the flame ionization detector (FID);
- the mass spectrometer (MS) detector.

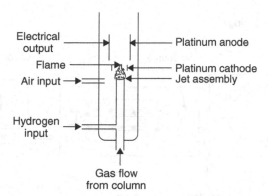

Figure 1.10 Schematic diagram of a flame-ionization detector. From Dean, J. R., *Bioavailability, Bioaccessibility and Mobility of Environmental Contaminants*, AnTS Series, Copyright 2007. © John Wiley & Sons, Limited. Reproduced with permission.

In the case of the FID (Figure 1.10) the exiting GC carrier gas stream, containing the separated organic compounds, passes through a (small) hydrogen flame that has a potential (>100 V) applied across it. As the organic compounds pass through the flame they become ionized, producing ions and electrons. It is the collection of these electrons that creates a small electric current that is amplified to produce a signal response proportional to the amount of organic compound. The FID is a very sensitive detector with a good linear response over a wide concentration range.

In the case of the mass spectrometer detector, compounds exiting the column are bombarded with electrons from a filament (electron impact or EI mode) (Figure 1.11) causing the compound to fragment with the production of charged species. It is these charged species which are then separated by a mass spectrometer (typically a quadrupole MS) based on their mass/charge ratio. Upon exiting the quadrupole the ions are detected by an electron multiplier tube which converts the positive compound ion (cation) into an electron, which is then multiplied and collected at an anode, resulting in a signal response which is proportional to the amount of organic compound. The MS can collect data in two formats: total ion current (TIC) (or full scan) mode and single (or selected) ion monitoring (SIM) mode.

SAQ 1.4

What is the difference in output between the TIC and SIM modes and how is it achieved?

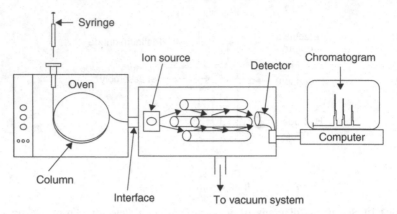

Figure 1.11 Schematic diagram of a capillary gas chromatography–mass spectrometry hyphenated system. From Dean, J. R., *Bioavailability, Bioaccessibility and Mobility of Environmental Contaminants*, AnTS Series, Copyright 2007. © John Wiley & Sons, Limited. Reproduced with permission.

1.5.2 High Performance Liquid Chromatography

In high performance liquid chromatography (HPLC) a mobile phase, into which the sample is introduced, passes through a column packed with micrometre-sized particles. HPLC allows rapid separation of complex mixtures of non-volatile compounds. A schematic diagram of an HPLC system is shown in Figure 1.12.

1.5.2.1 Mobile Phase for HPLC

The mobile phase for HPLC consists of an organic solvent (typically methanol or acetonitrile) and water (or buffer solution). The mobile phase is normally filtered (to remove particulates) and degassed (to remove air bubbles) prior to being pumped to the column by a reciprocating piston pump. The pumping system can operate in one of two modes allowing either isocratic or gradient elution of the non-volatile organic compounds. In the isocratic mode the same solvent mixture is used throughout the analysis while in the gradient elution mode the composition of the mobile phase is altered using a microprocessor-controlled gradient programmer, which mixes appropriate amounts of two different solvents to produce the required gradient. Gradient elution allows the separation of more complex organic compound mixtures rather than isocratic elution. Also, at the end of the gradient, elution time has to be allowed for a re-equilibration of the system to the initial mobile phase conditions. A typical gradient elution approach may consist of the following: start at an initial mobile phase composition of 30:70 vol/vol methanol:water for 2 min, then a linear gradient to 90:10 vol/vol methanol:water in 20 min, followed by a 'hold mobile phase composition' for 2 min. In order for the next sample to be introduced, the mobile phase composition

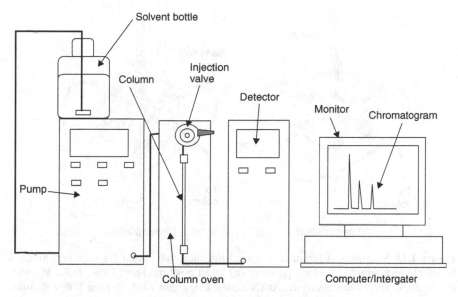

Figure 1.12 Schematic diagram of an isocratic high performance liquid chromatograph. Reproduced by permission of Mr E. Ludkin, Northumbria University, Newcastle, UK.

must return to the initial conditions, i.e. 30:70 vol/vol methanol:water prior to injection; this process is relatively rapid taking approximately 5–10 min.

1.5.2.2 Sample Introduction for HPLC

The most common method of sample introduction in HPLC is via a rotary 6-port valve, i.e. a Rheodyne® valve. A schematic diagram of a rotary 6-port valve is shown in Figure 1.13. Injection of a sample (or a standard) can be done either manually, by hand, or via a computer-controlled autosampler.

DQ 1.8

How might you manually inject a sample/standard into the chromatograph?

Answer

In the manual injection mode a sample/standard is introduced as follows:

- The syringe is filled (1.0 ml) with the sample/standard solution; this is achieved by inserting the needle of the syringe into the solution and slowly raising the plunger, taking care not to introduce any air bubbles.

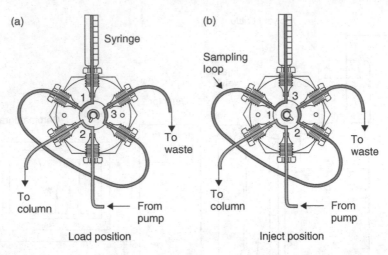

Figure 1.13 Schematic diagram of a typical injection valve used for high performance liquid chromatography: (a) load position; (b) inject position. From Dean, J. R., *Methods for Environmental Trace Analysis*, AnTS Series. Copyright 2003. © John Wiley & Sons, Limited. Reproduced with permission.

- The outside of the syringe is then wiped clean with a tissue.

- Then, the syringe is placed into the 6-port valve which is located in the 'load' position and the plunger depressed (but not all the way) to introduce the sample into an external loop of fixed volume (typically 5, 10 or 20 µl). While this is occurring the mobile phase passes through the 6-port valve to the column.

- Then, the 6-port valve is rotated into the 'inject' position. This causes the mobile phase to be diverted through the sample loop, thereby introducing a reproducible volume of the sample into the mobile phase.

The mobile phase transports the sample from the 6-port valve to the column.

1.5.2.3 HPLC Column

An HPLC column is made of stainless steel tubing with appropriate end fittings that allow coupling to connecting tubing (either stainless steel or PEEK). Typical column lengths vary between 1 and 25 cm (typically 25 cm) with an internal diameter of <1.0 mm to 4.6 mm (typically 4.6 mm). The stationary phase is bonded to silica particles (typically 3 or 5 µm diameter). Based on the composition of the mobile phase, described above, the chemically bonded stationary phase is typically C_{18} (also known as octadecylsilane (ODS)) (Figure 1.14). Other stationary

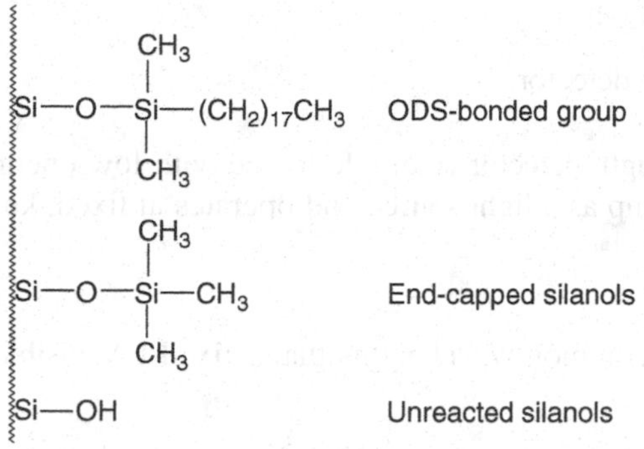

Figure 1.14 Silica particles coated with octadecylsilane (ODS) for reversed phase high performance liquid chromatography. From Dean, J. R., *Bioavailability, Bioaccessibility and Mobility of Environmental Contaminants*, AnTS Series, Copyright 2007. © John Wiley & Sons, Limited. Reproduced with permission.

phases include C_8, C_6, C_2 and C_1. The presence of unreacted silanol groups on the stationary phase can lead to detrimental compound separation.

SAQ 1.5

How might this detrimental separation be evident?

To compensate for these issues it is possible to obtain end-capped C_{18}; in this situation the silanol groups are blocked with C_1 entities. The column is often located within an oven which is used to stabilize peak elution. The temperature of the oven is maintained at a fixed temperature (e.g. typically in the range 23–35°C).

1.5.2.4 Detectors for HPLC

After HPLC separation the eluting compounds need to be detected. The most common detectors for HPLC are the universal detectors, as follows:

- the ultraviolet/visible detector (UV/visible);
- the mass spectrometry (MS) detector.

In the case of the UV/visible detectors they are widely used and have the advantages of versatility, sensitivity and stability. They are available in three forms:

- fixed wavelength;

- variable wavelength;
- as a diode array detector.

A fixed wavelength detector is simple to use with low operating costs. It contains a mercury lamp as a light source and operates at fixed, known wavelengths.

DQ 1.9

What are the common wavelengths that a fixed UV/visible detector can operate at?

Answer

Typically one of the following: 214, 254 or 280 nm.

Variable-wavelength detectors use a deuterium lamp and a continuously adjustable monochromator for wavelength coverage between 190 to 600 nm. The use of a diode array detector incorporates the advantage of multi-wavelength coverage with the ability to run a UV/visible spectrum for any compound detected. This 3-dimensional image of absorbance (i.e. the signal) versus compound elution time (i.e. the chromatogram) and a UV/visible spectrum is invaluable in chromatographic method development. The sensitivity of the UV/visible detector is influenced by the pathlength of the 'z-shaped' flow cell (typically 10 mm) which maximizes signal intensity (Figure 1.15).

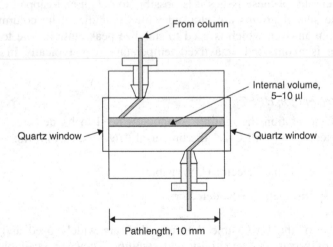

Figure 1.15 Schematic diagram of a UV/visible detector cell for high performance liquid chromatography. From Dean, J. R., *Bioavailability, Bioaccessibility and Mobility of Environmental Contaminants*, AnTS Series, Copyright 2007. © John Wiley & Sons, Limited. Reproduced with permission.

In the case of the mass spectrometry (MS) detector, compounds exiting the column are ionized at atmospheric pressure (i.e. external to the MS detector). The two major interfaces are:

- electrospray (ES) ionization;
- atmospheric pressure chemical ionization (APCI).

In ES ionization (Figure 1.16) the mobile phase is pumped through a stainless-steel capillary tube held at a potential of between 3 to 5 kV. This results in the mobile phase being sprayed from the exit of the capillary tube, producing highly charged solvent and solute ions in the form of droplets. Applying a continuous flow of nitrogen carrier gas allows the solvent to evaporate, leading to the formation of solute ions. These ions are introduced into the spectrometer via a 'sample-skimmer' arrangement. By allowing the formation of a potential gradient between the electrospray and the nozzle, the generated ions are introduced into the mass spectrometer.

In APCI the voltage (2.5–3.0 kV) is applied to a corona pin which is positioned in front of the stainless-steel capillary tubing through which the mobile phase from the HPLC passes (Figure 1.17). To assist the process the capillary tube is heated and surrounded by a coaxial flow of nitrogen gas. The interaction of the nitrogen gas and the mobile phase results in the formation of an aerosol which enters the corona discharge, producing sample ions. These ions are transported into the mass spectrometer in the same way as described above for ES. Using ES or APCI, organic compounds form singly charged ions by the loss or gain of a proton (hydrogen atom), i.e. $[M + 1]^+$ (typically basic compounds, e.g. amines)

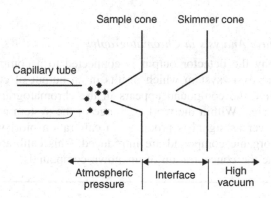

Figure 1.16 Schematic diagram of an electrospray ionization (ESI) source for HPLC–MS. From Dean, J. R., *Bioavailability, Bioaccessibility and Mobility of Environmental Contaminants*, AnTS Series, Copyright 2007. © John Wiley & Sons, Limited. Reproduced with permission.

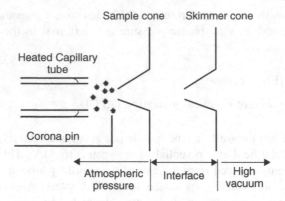

Figure 1.17 Schematic diagram of an atmospheric-pressure chemical ionization (APCI) source for HPLC–MS. From Dean, J. R., *Bioavailability, Bioaccessibility and Mobility of Environmental Contaminants*, AnTS Series, Copyright 2007. © John Wiley & Sons, Limited. Reproduced with permission.

or $[M - 1]^-$ (typically acidic compounds, e.g. carboxylic acids), where M is the molecular weight of the compound allowing the spectrometer to operate in either the positive ion mode or negative ion mode, respectively. Separation of the ions takes place in either a quadrupole mass spectrometer, ion-trap mass spectrometer or time-of-flight mass spectrometer. In order that both positive and negative ions can be detected in MS requires the use of an electron multiplier tube with a conversion dynode prior to the normal discrete dynode. The conversion dynode can be segmented: one segment coated with a material that is responsive to negative ions while a different segment is coated with a material that is responsive to positive ions.

1.5.2.5 Quantitative Analysis in Chromatography

In chromatography the detector output is connected to a computer-based data acquisition and analysis system which results in an output of compound retention time (the time the compound appears in the chromatogram) and its peak height and peak area. Within the working range of the system a linear response of concentration versus signal is produced (a calibration plot) when increasing amounts of the organic compound are introduced. This calibration plot is then used to determine the concentration of unknown compounds.

SAQ 1.6

The data in Table 1.3 have been obtained by a chromatography experiment for the determination of chlorobenzene. Plot the data on a calibration graph using 'Excel'.

Table 1.3 An example of how to record quantitative data from a chromatography experiment

Concentration (mg/l)	Signal
0	23
2.5	2345
5	4543
7.5	6324
10	8456
20	17 843

SAQ 1.7

If the signal response for an unknown sample, containing chlorobenzene, was 1234 what is the concentration of chlorobenzene in the sample?

Often in GC it is necessary to add an internal standard (a substance not present in the unknown sample, but with a similar chemical structure that elutes at a different time to other compounds present) to compensate for variation in injection volumes when introducing sample volumes in GC.

1.5.3 Sample Pre-Concentration Methods

Sometimes when the concentration of the organic compound in the sample extract is expected to be very low it is necessary to reduce the volume of organic solvent present in order to allow a pre-concentration effect. The most common approaches for solvent evaporation are gas blow-down, Kuderna–Danish evaporative concentration, the automated **eva**porative **c**oncentration **s**ystem (EVACS) or rotary evaporation. In all cases, the evaporation method is slow with the risk of contamination from the solvent, glassware and blow-down gas high. Sometimes the sample extract is taken to dryness and reconstituted in a very small volume (e.g. 100 µl) of organic solvent. Often vortex shaking is used to help re-solubilize the extract residue with the organic solvent. This approach is used when the lowest concentration levels are to be determined.

Gas blow-down The typical procedure for gas blow-down is carried out by blowing a stream of nitrogen over the surface of the solution, while gently warming the solution. A schematic diagram of the apparatus is shown in Figure 1.18. The sample is placed in an appropriately sized tube with a conical base. A gentle stream of nitrogen is directed towards the side of the tube so that it flows over the surface of the organic solvent extract which at the same time is being gently heated via a purposely designed aluminium heating block or water bath.

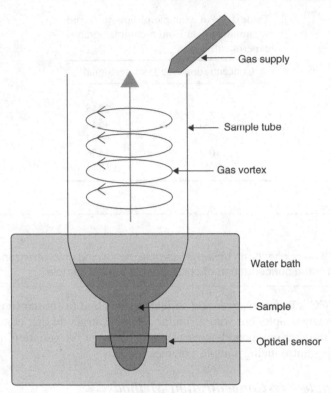

Figure 1.18 Schematic diagram of a typical gas 'blow-down' system (Tubovap™) used for the pre-concentration of compounds in organic solvents. From Dean, J. R., *Methods for Environmental Trace Analysis*, AnTS Series. Copyright 2003. © John Wiley & Sons, Limited. Reproduced with permission.

SAQ 1.8

How might you speed up the evaporation process?

1.5.3.1 Kuderna–Danish Evaporative Concentration

The Kuderna–Danish evaporative condenser [1] was developed in the laboratories of Julius Hyman and Company, Denver, Colorada, USA [2]. It consists of an evaporation flask (500 ml) connected at one end to a Snyder column and the other end to a concentrator tube (10 ml) (Figure 1.19). The sample containing organic solvent (200–300 ml) is placed in the apparatus, together with one or two boiling chips, and heated with a water bath. The temperature of the water bath should be maintained at 15–20°C above the boiling point of the organic solvent. The

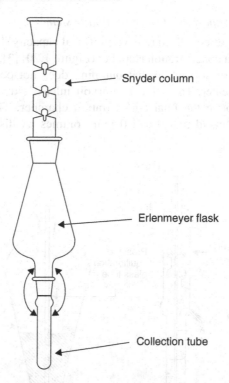

Snyder column

Erlenmeyer flask

Collection tube

Figure 1.19 Schematic diagram of the Kudema–Danish evaporative concentration condenser system. From Dean, J. R., *Methods for Environmental Trace Analysis*, AnTS Series. Copyright 2003. © John Wiley & Sons, Limited. Reproduced with permission.

positioning of the apparatus should allow partial immersion of the concentrator tube in the water bath but also allow the entire lower part of the evaporation flask to be bathed with hot vapour (steam). Solvent vapours then rise and condense within the Snyder column. Each stage of the Snyder column consists of a narrow opening covered by a loose-fitting glass insert. Sufficient pressure needs to be generated by the solvent vapours to force their way through the Snyder column. Initially, a large amount of condensation of these vapours returns to the bottom of the Kuderna–Danish apparatus. In addition to continually washing the organics from the sides of the evaporation flask, the returning condensate also contacts the rising vapours and assists in the process of recondensing volatile organics. This process of solvent distillation concentrates the sample to approximately 1–3 ml in 10–20 min. Escaping solvent vapours are recovered using a condenser and collection device. The major disadvantage of this method is that violent solvent eruptions can occur in the apparatus leading to sample losses. Micro-Snyder column systems can be used to reduce the solvent volume still further.

1.5.3.2 Automated Evaporative Concentration System

Solvent from a pressure-equalized reservoir (500 ml capacity) is introduced, under controlled flow, into a concentration chamber (Figure 1.20) [3]. Glass indentations regulate the boiling of solvent so that bumping does not occur. This reservoir is surrounded by a heater. The solvent reservoir inlet is situated under the level of the heater just above the final concentration chamber. The final concentration chamber is calibrated to 1.0 and 0.5 ml volumes. A distillation column is

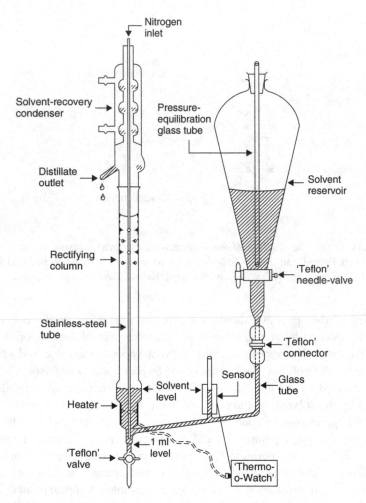

Figure 1.20 Schematic diagram of the automatic evaporative concentration system: ▨, solvent; ▢, vapour. Reprinted with permission from Ibrahim, E. A., Suffet, I. H. and Sakla, A. B., *Anal. Chem.*, **59**, 2091–2098 (1987). Copyright (1987) American Chemical Society.

connected to the concentration chamber. Located near the top of the column are four rows of glass indentations which serve to increase the surface area. Attached to the top of the column is a solvent recovery condenser with an outlet to collect and hence recover the solvent.

To start a sample, the apparatus is operated with 50 ml of high-purity solvent under steady uniform conditions at total reflux for 30 min to bring the system to equilibrium. Then the sample is introduced into the large reservoir either as a single volume or over several time intervals. (NOTE: A boiling point difference of approximately 50°C is required between solvent and analyte for the highest recoveries.) The temperature is maintained to allow controlled evaporation. For semi-volatile analytes this is typically at 5°C higher than the boiling point of the solvent. The distillate is withdrawn while keeping the reflux ratio as high as possible. During operation, a sensor monitors the level of liquid, allowing heating to be switched off or on automatically (when liquid is present the heat is on and vice versa). After evaporation of the sample below the sensor level, the heating is switched off. After 10 min the nitrogen flow is started to give a final concentration from 10 ml to 1 ml (or less). Mild heat can be applied according to the sensitivity of solvent and analyte to undergo thermal decomposition. When the liquid level drops below the tube, 'stripping' nearly stops. The tube is sealed at the bottom, so that the nitrogen is dispersed above the sample and the reduction of the volume becomes extremely slow. This prevents the sample from going to dryness even if left for hours. The sample is drained and the column is rinsed with two 0.5 ml aliquots of solvent. Further concentration can take place, if required.

1.5.3.3 Rotary Evaporation

Organic solvent is removed, under reduced pressure, by mechanically rotating a flask containing the sample in a controlled temperature water bath (Figure 1.21).

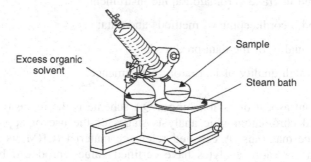

Figure 1.21 A typical rotary evaporation system used for the pre-concentration of compounds in organic solvents. From Dean, J. R., *Methods for Environmental Trace Analysis*, AnTS Series. Copyright 2003. © John Wiley & Sons, Limited. Reproduced with permission.

The waste solvent is condensed and collected for disposal. Problems can occur due to loss of volatile compounds, adsorption onto glassware, entrainment of compounds in the solvent vapour and the uncontrollable evaporation process. The sample residue is re-dissolved in the minimal quantity of solvent, assisted by vortex mixing.

1.6 Quality Assurance Aspects

Quality assurance is about designing laboratory protocols to obtain the correct result for the organic compounds being analysed. In analytical sciences, as we have seen in this chapter, the analytical process has several steps that include: sample collection, pre-treatment and storage which are then followed by extraction and chromatographic analysis.

While it is likely that the final errors in the data are greater from the sampling considerations rather than the laboratory-based aspects it is good practice to assess the laboratory quality assurance protocols. The most important terms in assessing these protocols are accuracy and precision. Accuracy is defined as the closeness of a determined value to its 'true' value, while precision is defined as the closeness of the determined values to each other. It is possible for the extraction and analysis of organic compounds from sample matrices to have combinations of accurate/inaccurate data alongside precise/imprecise data. The skill of the analytical scientist is to assess these variations such that accurate and precise data are obtained on laboratory samples.

The core components of a laboratory-based quality assurance scheme are to:

- select and validate appropriate methods of sample extraction;
- select and validate appropriate methods of chromatographic analysis;
- maintain and upgrade chromatographic instruments;
- ensure good recordkeeping of methods and data;
- ensure the quality of the data produced;
- maintain a high quality of laboratory performance.

An important aspect of establishing a QA scheme is the inclusion within the extraction and chromatographic analysis stages of the use of appropriate certified reference materials. A certified reference material (CRM) is a substance for which one or more analytes have certified values, produced by a technically valid procedure, accompanied with a traceable certificate and issued by a certifying body.

The major certifying bodies for CRMs are the National Institute for Standards and Technology (NIST) based in Washington DC, USA, the Community Bureau of Reference (known as BCR), Brussels, Belgium and the Laboratory of the Government Chemist (LGC), Teddington, U.K.

Other important procedures to build into any laboratory quality assurance protocols would include:

- Calibration with standards. A minimum number of standards should be used to generate the analytical calibration plot, e.g. 6 or 7. Daily verification of the working calibration plot should also be carried out using one or more standards within the linear working range while the selected standard should be 'sandwiched' between chromatographic runs of unknown sample extracts (typically every 10 unknown sample extracts).

- Analysis of reagent blanks. Analyse reagents whenever the batch is changed or a new reagent introduced. Introduce a minimum number of reagent blanks (typically 5% of the sample load) into the analytical protocol. This allows reagent purity to be assessed and, if necessary, controlled and also acts to assess the overall procedural blank.

- Analysis of precision. Repeat extractions from sub-samples, typically a minimum of three repeats required (ideally 7 repeat extractions of sub-samples should be used).

- Spiking studies on blanks and samples to establish recovery levels.

- Maintenance of control charts for standards and reagent blanks. The purpose is to assess the longer-term performance of the laboratory, instrument, operator or procedure, based on a statistical approach.

1.7 Health and Safety Considerations

All laboratory work must be carried out with due regard to Government legislation and employer guidelines. In the UK while the Health and Safety at Work Act (1974) provides the main framework for health and safety, it is the Control of Substances Hazardous to Health (COSHH) regulations of 1994 and 1996 that impose strict legal requirements for risk assessment of chemicals. Within the COSHH regulations the terms 'hazard' and 'risk' have very specific meanings; a hazardous substance is one that has the ability to cause harm whereas risk is about the likelihood that the substance may cause harm and is directly linked to the amount of chemical being used. For example, a large volume of flammable organic solvent has a greater risk than a small quantity of the same solvent.

All laboratories must operate a safety scheme. Your responsibility is to ensure that you comply with its operation to maintain safe working conditions for yourself and other people in the laboratory. A set of basic generic laboratory rules are described below:

(1) Always wear appropriate protective clothing, a clean laboratory coat, safety glasses/goggles and appropriate footwear. It may be necessary to wear protective gloves when handling certain chemicals.

(2) You must never eat or drink in the laboratory.

(3) You must never work alone in a laboratory.

(4) You must ensure that you are familiar with the fire regulations in your laboratory and building.

(5) You should be aware of accident/emergency procedures in your laboratory and building.

(6) Always use appropriate devices for transferring liquid, e.g. a pipette, syringe, etc.

(7) Always use a fume cupboard for work with hazardous (including volatile, flammable) chemicals.

(8) Always clear up any spillages as they occur.

(9) It is advisable to plan your work in advance; work in a logical and methodical manner.

Summary

This chapter initially summarizes the important considerations necessary in planning the whole analytical protocol, including pre-sampling, sampling, extraction and analysis for organic compounds from solid, aqueous and air samples. The main practical aspects of undertaking gas chromatography and high performance liquid chromatography are described as well as sample extract pre-concentration approaches that may be necessary for pre-analysis. Finally, a general description of quality assurance in an analytical laboratory is described, followed by the important health and safety considerations.

References

1. Karasek, F. W., Clement, R.E. and Sweetman, J.A., *Anal. Chem.*, **53**, 1050A–1058A (1981).
2. Gunther, F. A., Blinn, R. C., Kolbezen, M. J. and Barkley, J. H., *Anal. Chem.*, **23**, 1835–1842 (1951).
3. Ibrahim, E. A., Suffet, I. H. and Sakla, A. B., *Anal. Chem.*, **59**, 2091–2098 (1987).

AQUEOUS SAMPLES

Chapter 2

Classical Approaches for Aqueous Extraction

2.1 Introduction

The most common approach for the extraction of compounds from aqueous samples is liquid–liquid extraction (LLE). In addition, a brief description of the purge and trap technique which is used for volatile organic compounds in aqueous samples is also described.

2.2 Liquid–Liquid Extraction

The principal of liquid–liquid extraction is that a sample is distributed or partitioned between two immiscible liquids or phases in which the compound and

Extraction Techniques in Analytical Sciences   John R. Dean
© 2009 John Wiley & Sons, Ltd

matrix have different solubilities. Normally, one phase is aqueous (often the denser or heavier phase) and the other phase is an organic solvent (the lighter phase). The basis of the extraction process is that the more polar hydrophilic compounds prefer the aqueous (polar) phase and the more non-polar hydrophobic compounds prefer the organic solvent.

DQ 2.1

If the method of separation to be used is reversed phase high performance liquid chromatography (HPLC), in which phase are the target organic compounds best isolated?

Answer

If the method of separation to be used is reversed phase high performance liquid chromatography (HPLC), then the target organic compounds are best isolated in the aqueous phase so that they can be directly injected into the HPLC system.

Alternatively, if the target organic compounds are to be analysed by gas chromatography they are best isolated in an organic solvent. The compounds in the organic solvent (for GC) can be analysed directly or pre-concentrated further using, for example, solvent evaporation (see Chapter 1), while compounds in the aqueous phase (for HPLC) can be analysed directly or pre-concentrated further using, for example, solid phase extraction (see Chapter 3). The main advantages of LLE are its wide applicability, availability of high purity organic solvents and the use of low-cost apparatus (e.g. a separating funnel).

2.2.1 Theory of Liquid–Liquid Extraction

Two terms are used to describe the distribution of a compound between two immiscible solvents, namely the distribution coefficient and the distribution ratio.

The distribution coefficient is an equilibrium constant that describes the distribution of a compound, X, between two immiscible solvents, e.g. an aqueous phase and an organic phase. For example, an equilibrium can be obtained by shaking the aqueous phase containing the compound, X, with an organic phase, such as hexane. This process can be written as an equation:

$$X(aq) \longleftrightarrow X(org) \tag{2.1}$$

where (aq) and (org) are the aqueous and organic phases, respectively. The ratio of the activities of X in the two solvents is constant and can be represented by:

$$K_d = [X]_{org}/[X]_{aq} \tag{2.2}$$

where K_d is the distribution coefficient. While the numerical value of K_d provides a useful constant value, at a particular temperature, the activity coefficients are neither known or easily measured. A more useful expression is the fraction of compound extracted (E), often expressed as a percentage:

$$E = C_o V_o / (C_o V_o + C_{aq} V_{aq}) \tag{2.3}$$

or:

$$E = K_d V / (1 + K_d V) \tag{2.4}$$

where C_o and C_{aq} are the concentrations of the compound in the organic phase and aqueous phases, respectively, V_o and V_{aq} are the volumes of the organic and aqueous phases, respectively, and V is the phase ratio, V_o / V_{aq}.

For one-step liquid–liquid extractions, K_d must be large, i.e. >10, for quantitative recovery ($>99\%$) of the compound in one of the phases, e.g. the organic solvent. This is a consequence of the phase ratio, V, which must be maintained within a practical range of values: $0.1 < V < 10$ (Equation (2.4)). Typically, two or three repeat extractions are required with fresh organic solvent to achieve quantitative recoveries. Equation (2.5) is used to determine the amount of compound extracted after successive multiple extractions:

$$E = 1 - [1/(1 + K_d V)]^n \tag{2.5}$$

where n is the number of extractions.

SAQ 2.1

If the volumes of the two phases are equal ($V = 1$) and $K_d = 3$ for a compound, then how many extractions would be required to achieve $>99\%$ recovery?

It can be the situation that the actual chemical form of the compound in the aqueous and organic phases is not known, e.g. a variation in pH would have a significant effect on a weak acid or base. In this case the distribution ratio, D, is used:

$$D = \frac{\text{concentration of X in all chemical forms in the organic phase}}{\text{concentration of X in all chemical forms in the aqueous phase}} \tag{2.6}$$

(Note: for simple systems, when no chemical dissociation occurs, the distribution ratio is identical to the distribution coefficient.)

2.2.2 Selection of Solvents

The selectivity and efficiency of LLE is critically governed by the choice of the two immiscible solvents. Often the organic solvent for LLE is chosen because of its:

- Low solubility in the aqueous phase (typically <10%).

- High volatility for solvent evaporation in the concentration stage (see Chapter 1, Section 1.5.3).

- High purity (directly linked to the solvent evaporation process, described above) which could pre-concentrate any impurities within the solvent.

- Compatibility with the choice of chromatographic analysis. For example, do not use chlorinated solvents, such as, dichloromethane, if the method of analysis is GC–ECD (Chapter 1, Section 1.5.2) or strongly UV-absorbing solvents if using HPLC–UV (Chapter 1, Section 1.5.2).

- Polarity and hydrogen-bonding properties that can enhance compound recovery in the organic phase, i.e. increase the value of K_d (Equation 2.2).

The equilibrium process (K_d) can be influenced by several factors that include adjustment of pH to prevent ionization of acids or bases, by formation of ion-pairs with ionizable compounds, by formation of hydrophobic complexes with metal ions or by adding neutral salts to the aqueous phase to reduce the solubility of the compound ('salting out'). Examples of the choice of solvents for LLE are shown in Table 2.1 [1].

2.2.3 Solvent Extraction

Two distinct approaches for LLE are possible, i.e. discontinuous LLE, where equilibrium is established between two immiscible phases, or continuous LLE, where equilibrium may not be reached.

In discontinuous extraction the most common approach uses a separating funnel (Figure 2.1). The aqueous sample (1 l, at a specified pH) is introduced into a large

Table 2.1 Solvents for LLE [1]

Aqueous solvents	Water-immiscible organic solvents
Water	Hexane, isooctane, petroleum ether (or other aliphatic hydrocarbons)
Acidic solution	Diethylether
Basic solution	Dichloromethane
High salt ('salting-out' effect)	Chloroform
Complexing agents (ion pairing, chelating and chiral)	Ethyl acetate
Any two (or more) of the above	Aliphatic ketones (C6 and above)
	Aliphatic alcohols (C6 and above)
	Toluene, xylenes (UV absorbance)
	Any two (or more) of the above

Figure 2.1 A separating funnel. From Dean, J. R., *Extraction Methods for Environmental Analysis*, Copyright 1998. © John Wiley & Sons, Limited. Reproduced with permission.

separating funnel (21 capacity with a Teflon stopcock) and 60 ml of a suitable organic solvent, e.g. dichloromethane, is added. A stopper is then placed into the top of the separating funnel and the separating funnel is then shaken manually. By placing the stoppered end of the separating funnel into the palm of the hand an inversion of the funnel can take place. This process is repeated for approximately 1–2 min (inverting the separating funnel approximately 5–6 times).

SAQ 2.2

Why should the stopcock be opened in between each inversion of the separating funnel?

The process can also be automated by using a mechanical 'bed-shaker'. The shaking process allows thorough interspersion between the two immiscible solvents, thereby maximizing the contact between the two solvent phases and hence assisting mass transfer, and allowing efficient partitioning to occur. After a suitable resting period (approximately 5 min) the organic solvent is collected by opening the stopcock and carefully running out the lower phase (assuming this to be the organic phase) and quantitatively transferred to a volumetric flask. Fresh organic solvent is then added to the separating funnel and the process repeated again. This should be done at least three times in total. The three organic extracts should be combined, ready for concentration (see Chapter 1, Section 1.5.3).

In some cases where the kinetics of the extraction process are slow, such that the equilibrium of the compound between the aqueous and organic phases is poor, i.e. K_d is very small, then continuous LLE can be used. This approach can

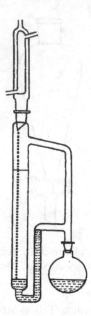

Figure 2.2 Continuous liquid–liquid extraction (organic solvent heavier than water). From Dean, J. R., *Extraction Methods for Environmental Analysis*, Copyright 1998. © John Wiley & Sons, Limited. Reproduced with permission.

also be used for large volumes of aqueous sample. In this situation, fresh organic solvent is boiled, condensed and allocated to percolate repetitively through the compound-containing aqueous sample. Two common versions of continuous liquid extractors are available, using either lighter-than or heavier-than water organic solvents (Figure 2.2). Extractions usually take several hours, but do provide concentration factors of up to $\times 10^5$. Obviously several systems can be operated unattended and in series, allowing multiple samples to be extracted. Typically, a 1 l sample, pH adjusted if necessary, is added to the continuous extractor. Then organic solvent, e.g. dichloromethane (in the case of a system in which the solvent has a greater density than the sample), of volume 300–500 ml, is added to the distilling flask together with several boiling chips. The solvent is then boiled, using a water bath, and the extraction process continues for 18–24 h. After completion of the extraction process, and allowing for suffcient cooling time, the boiling flask is detached and solvent evaporation can then occur (see Chapter 1, Section 1.5.3).

2.2.4 Problems with the LLE Process

Practical problems with LLE can occur and include emulsion formation. The latter can occur particularly for samples that contain surfactants or fatty materials.

DQ 2.2

In LLE, what is an emulsion?

Answer

An emulsion appears as a 'milky white' colouration within the separating funnel with no distinct boundary between the aqueous and organic phases.

DQ 2.3

How can an emulsion be remedied?

Answer

The remedy is to disrupt or 'break-up' the emulsion by:

- centrifugation of the mixture;
- filtration through a glass wool plug or phase separation paper;
- heating (e.g. place in an oven) or cooling (e.g. place in a refrigerator) the separating funnel;
- 'salting-out' by addition of sodium chloride salt to the aqueous phase;
- addition of a small amount of a different organic solvent.

2.3 Purge and Trap for Volatile Organics in Aqueous Samples

Purge and trap is a widely applicable technique for the extraction of volatile organic compounds (VOCs) from aqueous samples, followed by direct transfer and introduction into the injection port of a gas chromatograph. An aqueous sample (e.g. 5 ml) is placed into a glass 'sparging' vessel (Figure 2.3). The sample is then 'purged' with high-purity nitrogen at a flow rate of 40–50 ml min^{-1} for 10–12 min. The recovered VOCs are then transferred to a trap, e.g. Tenax, at ambient temperature (see also Chapter 11). Desorption of the VOCs from the trap takes place by rapidly heating the trap (180–250°C) and back-flushing off the VOCs, in a stream of nitrogen gas, to the chromatograph. The rapid desorption from the trap occurs within 2–4 min and with a nitrogen flow rate of 1–2 ml min^{-1} and allows the VOCs to be desorbed in a sharp 'plug'. The VOCs are maintained in the gaseous form by ensuring that the transfer line from the trap to the chromatograph is independently heated (e.g. 225°C). The heated

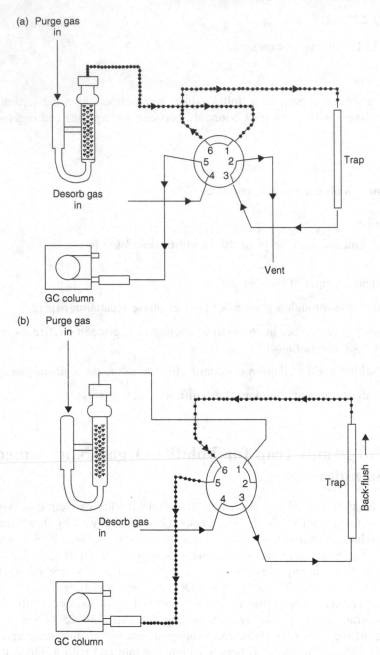

Figure 2.3 Illustrations of typical layouts for purge-and-trap extraction of volatile organic compounds from aqueous samples: (a) in 'purge mode'; (b) in 'desorb mode' (→ indicates sample pathway). From Dean, J. R., *Methods for Environmental Trace Analysis*, AnTS Series. Copyright 2003. © John Wiley & Sons, Limited. Reproduced with permission.

transfer line is introduced directly into the injection port of the chromatograph. At the end of each extraction, the trap can be 'baked out' by heating to 230°C for 8 min to remove any residual contaminants.

Summary

The classical approach for recovering organic compounds from aqueous samples, namely liquid–liquid extraction, is discussed in this chapter. As well as providing the necessary background to the approach the important practical aspects of the technique are described. For completeness, the alternative approach for volatile organic compounds in aqueous samples, i.e. purge and trap, is described.

References

1. Majors, R. E., *LC–GC Europe*, **22**(3), 143–147 (2009).

Chapter 3
Solid Phase Extraction

Learning Objectives

- To be aware of approaches for performing solid phase extraction of organic compounds from aqueous samples.
- To be aware of the important variables in performing solid phase extraction.
- To be able to select the most appropriate sorbent for solid phase extraction.
- To understand the practical aspects of solid phase extraction.
- To know the principle of operation of solid phase extraction.
- To appreciate the practical difficulties that can arise in performing solid phase extraction and their remedies.
- To be aware of the potential of solid phase extraction for on-line operation.
- To be aware of the practical applications of solid phase extraction.

3.1 Introduction

Solid phase extraction (SPE) is a popular sample preparation method used for isolation, enrichment and/or clean-up of components of interest from aqueous samples. SPE normally involves bringing an aqueous sample into contact with a solid phase or sorbent whereby the compound is selectively adsorbed onto the surface of the solid phase prior to elution [1]. The solid phase sorbent is usually packed into small tubes or cartridges (compare with a liquid chromatography column in Chapter 1, Section 1.5.2). Recently many developments in SPE technology have taken place including new formats (e.g. discs, pipette tips and 96-well plates), new sorbents (e.g. silica or polymer-based media and mixed-mode

Extraction Techniques in Analytical Sciences   John R. Dean
© 2009 John Wiley & Sons, Ltd

media) and the development of automated and on-line systems [2]. Whichever design is used the sample-containing solvent is forced by pressure or vacuum through the sorbent. By careful selection of the sorbent, the organic compound should be retained by the sorbent in preference to other extraneous material present in the sample. This extraneous material can be washed from the sorbent by the passing of an appropriate solvent. Subsequently the compound of interest can then be eluted from the sorbent using a suitable solvent. This solvent is then collected for analysis. Further sample clean-up or preconcentration can be carried out, if desired.

DQ 3.1

What are the important variables in SPE?

Answer

The important variables in SPE are the choice of sorbent and the solvent system used for effective pre-concentration and/or clean-up of the compound in the sample.

The process of SPE should allow more affective detection and identification of the compounds in aqueous samples.

3.2 Types of SPE Media (Sorbent)

Generally SPE sorbents can be divided into three classes, i.e. normal phase, reversed phase and ion exchange. The most common sorbents are based on silica particles (irregular shaped particles with a particle diameter between 30 and 60 µm) to which functional groups are bonded to surface silanol groups to alter their retentive properties (it should be noted that unmodified silica is sometimes used). The bonding of the functional groups is not always complete and so unreacted silanol groups remain. These unreacted sites are polar, acidic sites and can make the interaction with compounds more complex. To reduce the occurrence of these polar sites, some SPE media are 'end-capped'.

SAQ 3.1

What is end-capping?

It is the nature of the functional groups that determines the classification of the sorbent. In addition to silica some other common sorbents are based on florisil, alumina and macroreticular polymers.

Normal phase sorbents have polar functional groups e.g. cyano, amino and diol (also included in this category is unmodified silica). The polar nature of these sorbents means that it is more likely that polar compounds, e.g. phenol, will be retained. In contrast, reversed phase sorbents have non-polar functional groups, e.g. octadecyl, octyl and methyl, and conversely are more likely to retain non-polar compounds, e.g. polycyclic aromatic hydrocarbons. Ion exchange sorbents have either cationic or anionic functional groups and when in the ionized form attract compounds of the opposite charge. A cation exchange phase, such as benzenesulfonic acid, will extract a compound with a positive charge (e.g. phenoxyacid herbicides) and vice versa. A summary of commercially available silica-bonded sorbents is given in Table 3.1.

3.2.1 Multimodal and Mixed-Phase Extractions

SPE normally takes place using one device (e.g. a cartridge) with a single sorbent (e.g. C18). However, if more than one type or class of compound is present in the aqueous sample or if additional selectivity is needed to isolate a specific compound, then multimodal SPE can be used. Multimodal SPE can be done in one of two ways: either by connecting two alternate phase SPE cartridges in series or by having two different functional group sorbents present within one cartridge.

In each case it would be possible, for example, to separate a hydrophobic organic compound and inorganic cations using multimodal SPE.

DQ 3.2

By consulting Table 3.1 which two SPE sorbents would you suggest for the multimodal retention of a hydrophobic organic compound and inorganic cations?

Answer

The concentration of a hydrophobic organic compound could be done using a reversed phase sorbent, e.g. C18 whereas the inorganic cations could be done using a strong cation cartridge (SCX).

3.2.2 Molecularly Imprinted Polymers (MIPs)

In recent years, molecularly imprinted polymers (MIPs) have been developed to use as sorbents in SPE. The use of MIPs has been shown to be more selective for the extraction of target compounds from complex matrices such as aqueous samples or organic extracts, as they are engineered cross-linked polymers synthesized with artificial generated recognition sites able to specifically retain a target molecule in preference to other closely related compounds. In addition,

Table 3.1 Some commonly available silica-bonded sorbents (adapted from Moors *et al.* [1]). Reproduced by permission of the International Union of Pure and Applied Chemistry from Moors *et al.*, *Pure Appl. Chem.*, **66**, 277–304 (1994)

Phase	Bonded moiety
Nonpolar phases	
C1, methyl	$Si-CH_3$
C8, octyl	$Si-(CH_2)_7-CH_3$
C18, octadecyl	$Si-(CH_2)_{17}-CH_3$
Polar phases	
Si, silica	$Si-OH$
CN, cyanopropyl	$Si-CH_2-CH_2-CH_2-CN$
2OH, diol	$Si-CH_2-CH_2-CH_2-O-CH_2-CHOH-CH_2OH$
Ion-exchange phases	
SCX, benzenesulfonic acid	$Si-CH_2-CH_2-CH_2-C_6H_4-SO_3{}^-$
DEA, diethylammoniopropyl tertiary amine	$Si-CH_2-CH_2-CH_2-NH^+-(CH_2-CH_3)_2$
SAX, trimethylammoniopropyl quaternary amine	$Si-CH_2-CH_2-CH_2-N^+-(CH_3)_3$

MIPs offer more flexibility in analytical methods as they are stable to extremes of pH, organic solvents and temperature [3]. The extraction procedures using MIPs are identical to other SPE media, i.e. the stages of wetting and conditioning of sorbent, sample loading, washing and compound elution have to be carried out. Hence, a careful selection of the most appropriate solvent to be applied in each step is important in order to separate the compound selectively. Numerous studies of the applications of MIPs since the year 2000 have been reviewed [4]. These studies deal with the extraction of organic compounds from various matrices, including water, sediment, soil, plants, body fluids, diesel fuel, gasoline and foods.

3.3 SPE Formats and Apparatus

The design of the SPE device can vary, with each design having its own advantages related to the number of samples to be processed and the nature of the sample and its volume. The most common arrangement is the syringe barrel or cartridge. The cartridge itself is usually made of polypropylene (although glass and polytetrafluorethylene, PTFE, are also available) with a wide entrance, through which the sample is introduced, and a narrow exit (male luer tip). The appropriate sorbent material, ranging in mass from 50 mg to 10 g, is positioned between two frits, at the base (exit) of the cartridge, which act to both retain the sorbent material and to filter out particulate matter. Typically the frit is made from polyethylene with a 20 μm pore size.

Solvent flow through a single cartridge is typically done using a side-arm flask apparatus (Figure 3.1), whereas multiple cartridges can be simultaneously processed (from 8 to 30 cartridges) using a commercially available vacuum manifold (Figure 3.2). In both cases a vacuum pump is required to affect the movement of solvent/sample through the sorbent.

SAQ 3.2

How might a manual SPE procedure, i.e. one with no vacuum pump, be carried out?

The most distinctly different approach to SPE is the use of a disc, not unlike a common filter paper. This SPE disc format is referred to by its trade name of 'Empore' discs. The 5–10 μm sorbent particles are intertwined with fine threads of PTFE which results in a disc approximately 0.5 mm thick and a diameter in the range 47 to 70 mm. Empore discs are placed in a typical solvent filtration system and a vacuum applied to force the solvent containing the sample through (Figure 3.3). To minimize dilution effects that can occur it is necessary to introduce a test tube into the filter flask to collect the final extract. Manifolds are commercially available for multiple sample extraction using Empore discs.

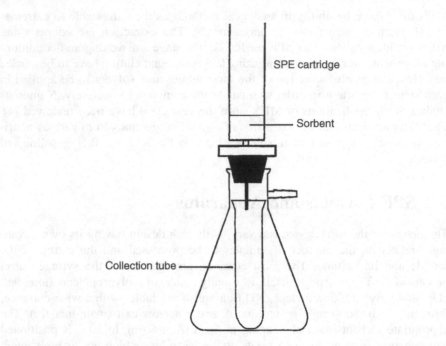

Figure 3.1 Solid phase extraction using a cartridge and a single side-arm flask apparatus. From Dean, J. R., *Extraction Methods for Environmental Analysis*, Copyright 1998. © John Wiley & Sons, Limited. Reproduced with permission.

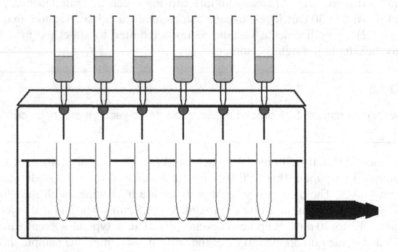

Figure 3.2 Vacuum manifold for solid phase extraction of multiple cartridges. For example, 10 SPE cartridges; 5 shown in the cross-section and another 5 located behind.

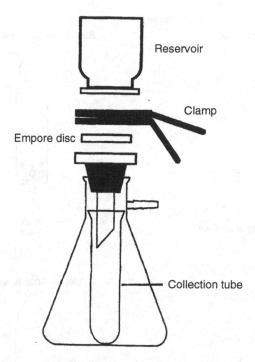

Figure 3.3 Solid phase extraction using an 'Empore' disc and a single side-arm flask apparatus. From Dean, J. R., *Extraction Methods for Environmental Analysis*, Copyright 1998. © John Wiley & Sons, Limited. Reproduced with permission.

Both the cartridge and disc formats have their inherent advantages and limitations.

SAQ 3.3

What are the advantages and limitations of an SPE disc?

3.4 Method of SPE Operation

Irrespective of the SPE format the method of operation is the same and can be divided into five steps (Figure 3.4) [1]. Each step is characterized by the nature and type of solvent used which in turn is dependent upon the characteristics of the sorbent and the sample.

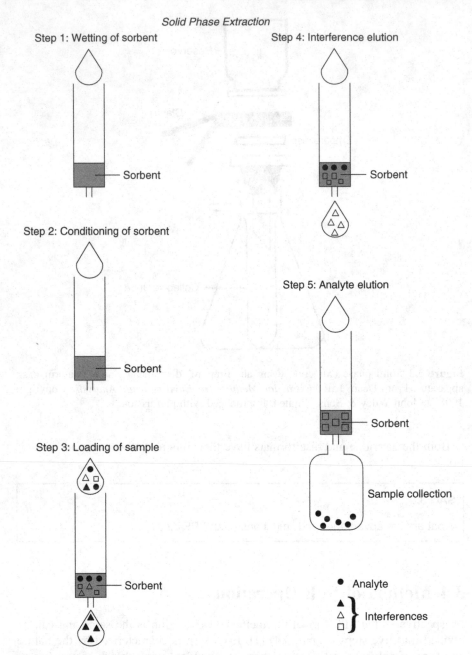

Figure 3.4 The five stages of operation of solid phase extraction. From Dean, J. R., *Extraction Methods for Environmental Analysis*, Copyright 1998. © John Wiley & Sons, Limited. Reproduced with permission.

DQ 3.3

What are the five stages of SPE operation?

Answer

The five stages are as follows:

- wetting the sorbent;
- conditioning of the sorbent;
- loading of the aqueous sample;
- rinsing or washing the sorbent to elute extraneous material;
- elution of the compound of interest.

Wetting the sorbent allows the bonded alkyl chains, which are twisted and collapsed on the surface of the silica, to be solvated so that they 'spread open' to form a 'bristle'. This ensures good contact between the compound and the sorbent in the adsorption stage. It is also important that the sorbent remains wet in the following two stages or poor recoveries can result. This is followed by conditioning of the sorbent in which solvent or buffer, similar in composition to the aqueous sample that is to be extracted, is pulled through the sorbent. (For aqueous samples this might be deionized, distilled water.) This is followed by sample loading where the sample is forced through the sorbent material by suction, a vacuum manifold or a plunger. By careful choice of the sorbent, it is anticipated that the compound of interest will be retained by the sorbent in preference to extraneous material and other related compounds of interest that may be present in the sample. Obviously this ideal situation does not always occur and compounds with similar structures will undoubtedly also be retained. This process is followed by washing with a suitable solvent that allows unwanted extraneous material to be removed without influencing the elution of the compound of interest. This stage is obviously the key to the whole process and is dependent upon the compound of interest and its interaction with the sorbent material and the choice of solvent to be used. Finally the compound of interest is eluted from the sorbent using the minimum amount of solvent to affect quantitative release. By careful control of the amount of solvent used in the elution stage and the sample volume initially introduced onto the sorbent a pre-concentration of the compound of interest can be affected. Successful SPE obviously requires careful consideration of the nature of the SPE sorbent, the solvent systems to be used and their influence on the compound of interest. In addition, it may be that it is not a single compound that you are seeking to pre-concentrate but a range of compounds. If they have similar chemical structures then a method can be successfully developed to extract these 'multiple-compounds'. While this method development may seem to be laborious and extremely time-consuming it should be remembered that multiple vacuum manifolds are commercially available as are robotic systems that can carry out the entire SPE process. Once developed, the SPE method can then be used to process large quantities of sample with good precision.

3.5 Solvent Selection

The choice of solvent directly influences the retention of the compound on the sorbent and its subsequent elution, whereas the solvent polarity determines the solvent strength (or ability to elute the compound from the sorbent in a smaller volume than a weaker solvent). The solvent strengths for normal phase and reversed phase sorbents are shown in Table 3.2. Obviously this is the ideal. In some situations it may be that no individual solvent will perform its function adequately and so it is necessary to resort to mixed solvent systems. It should also be noted that for a normal phase solvent, both solvent polarity and solvent strength are coincident whereas this is not the case for a reversed phase sorbent. In practice, however, the solvents normally used for reversed phase sorbents are restricted to water, methanol, isopropyl alcohol and acetonitrile. For ion exchange sorbents, solvent strength is not the main effect.

Table 3.2   Solvent strengths for normal and reversed phase sorbents. From Dean, J. R., *Extraction Methods for Environmental Analysis*, Copyright 1998. © John Wiley & Sons, Limited. Reproduced with permission

Solvent strength for normal phase sorbents		Solvent strength for reversed phase sorbents
Weakest	Hexane	**Strongest**
	Iso-octane	
	Toluene	
	Chloroform	
	Dichloromethane	
	Tetrahydrofuran	
	Ethyl ether	
	Ethyl acetate	
	Acetone	
	Acetonitrile	
	Isopropyl alcohol	
Strongest	Methanol	
	Water	**Weakest**

DQ 3.4

What do you think might be the key influencing parameters for ion exchange sorbents?

Answer

The main influencing parameters governing compound retention on the sorbent and its subsequent elution are pH and ionic strength.

As with the choice of sorbent some preliminary work is required to affect the best solvents to be used.

SAQ 3.4

Using a reversed phase sorbent (e.g. C18) as an example, what is the general methodology to be followed for SPE?

3.6 Factors Affecting SPE

While the choice of SPE sorbent is highly dependent upon the compound of interest and the sorbent system to be used, certain other parameters can influence the effectiveness of the SPE methodology. Obviously the number of active sites available on the sorbent cannot be exceeded by the number of molecules of compound or otherwise breakthrough will occur. Therefore, it is important to assess the capacity of the SPE cartridge or disc for its intended application. In addition, the flow rate of the sample through the sorbent is important; too fast a flow and this will allow minimal time for compound–sorbent interaction. This must be carefully balanced against the need to pass the entire sample through the cartridge or disc. It is normal therefore for an SPE cartridge to operate with a flow rate of 3–10 ml min^{-1} whereas 10–100 ml min^{-1} is typical for the disc format.

Once the compound of interest has been adsorbed by the sorbent, it may be necessary to wash the sorbent of extraneous matrix components prior to elution of the compound. The choice of solvent is critical in this stage, as has been discussed previously. For the elution stage it is important to consider the volume of solvent to be used (as well as its nature). For quantitative analysis, by, for example, HPLC or GC, two factors are important: (a) pre-concentration of the compound of interest from a relatively large volume of sample to a small extract volume and (b) clean-up of the sample matrix to produce a particle-free and chromatographically clean extract. All of these factors require some method development, either using a trial-and-error approach or by consultation with existing literature. It is probable that both are required in practice.

3.7 Selected Methods of Analysis for SPE

The general methodology to be followed for off-line SPE will be described using selected literature examples with emphasis on normal phase, reversed phase and ion exchange systems.

3.7.1 Applications of Normal Phase SPE

Normal phase (NP) SPE refers to the sorption of the functional groups of the compound (solute) from a non-polar solvent to the polar surface of the stationary phase such as silica gel, Florisil ($MgSiO_3$) and alumina (Al_2O_3). The mechanism of sorption involves polar interactions such as hydrogen bonding, dipole–dipole interactions, π–π interactions and induced dipole–dipole interactions. To achieve retention, the interaction between the solute and the stationary phase must dominate. Selected applications of NP SPE involving removal of organic compounds from non-polar solvents have been reported and are described in the following.

3.7.1.1 Analysis of Chlorinated Pesticides in Fish Extracts [5]

Chlorinated pesticides are known as environmentally persistent organic pollu-tants. They tend to accumulate in biological tissues due to their lipophilicity and generate adverse effects to living organisms. SPE was used as a method for sample clean-up of the fish extract prior to quantitative analysis of the pesticides.

Samples: Fish tissue samples were homogenized and extracted by ultrasonic agitation and lipids in the extract were eliminated by 'freezing-lipid' filtration; the sample extract was then concentrated to 1 ml by a rotary evaporator under a nitrogen atmosphere.

Compounds: 24 Chlorinated pesticides (examples of compounds are shown in Figure 3.5).

Sorbent: Florisil SPE cartridge, 2 g.

Wetting/Conditioning: The cartridge is cleaned with 12 ml of hexane and air dried for 1 min, followed by conditioning with 5 ml of hexane.

Loading: 1 ml of the sample extract was loaded onto the cartridge.

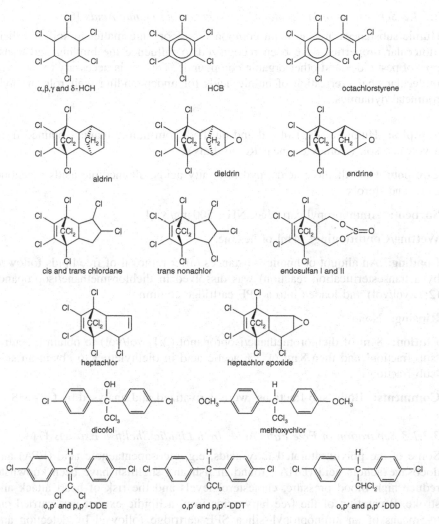

Figure 3.5 Structure of the chlorinated pesticides [5]. Reprinted from *J. Chromatogr., A*, **1038**(1/2), Hong *et al.*, 'Rapid determination of chlorinated pesticides in fish by freezing-lipid filtration, solid-phase extraction and gas chromatography–mass spectrometry', 27–35, Copyright (2004) with permission from Elsevier.

Rinsing: None.

Elution: 13 ml of acetone/*n*-hexane (1:9, vol/vol), at a flow rate of 1 ml min^{-1}.

Comments: The extract was then concentrated at 45°C with a nitrogen stream until dryness and an internal standard added prior to GC–MS analysis.

3.7.1.2 Separation of Molecular Constituents from Humic Acids [6]

Humic substances are the main components of organic matter in soil and their molecular properties have been recognized to influence the binding and transport of pesticides and other organic compounds. Thus, it is necessary to improve molecular characterization of humic acids for understanding their role in environmetal dynamics.

Samples: Humic acids, isolated and purified from humic matter obtained from a volcanic soil (from Vico, near Rome, Italy).

Compounds: Alkanoic acids, hydroxy fatty acids, alkanedioic acids, phenolic acids and sterols.

Sorbent: Aminopropyl cartridge, NH_2, 500 mg/3 ml.

Wetting/Conditioning: 4 ml of hexane.

Loading: An aliquot of humic substances (after removal of free lipids followed by a transesterification reaction) was dissolved in dichloromethane/isopropanol (2:1, vol/vol) and loaded into a SPE cartridge column.

Rinsing: None.

Elution: 8 ml of dichloromethane/isopropanol (2:1, vol/vol) to obtain a neutral 'sub-fraction' and then 8 ml of 2% acetic acid in diethylether to obtain an acid 'sub-fraction'.

Comments: Both 'sub-fractions' were derivatized and analysed by GC–MS.

3.7.1.3 Separation of Free Fatty Acids from Lipidic Shellfish Extracts [7]

Some of the polyunsaturated fatty acids, e.g. eicosapentaenoic acid (EPA) and docosahexaenoic acid (DHA), found in fish and shellfish have been known to reduce high blood pressure, cholesterol levels and the risk of heart attack and stroke. Separation of the free fatty acids from a lipidic extract was carried out by means of an aminopropyl–silica SPE cartridge followed by detection and quantification using LC–MS.

Sample: Lipidic shellfish extract.

Compounds: Free fatty acids.

Sorbent: Aminopropyl–silica cartridge (Discovery DSC-NH_2, 100 mg, 1 ml).

Wetting/Conditioning: 3 ml of chloroform.

Loading: 0.5 ml of a lipidic shellfish extract was loaded onto a cartridge.

Rinsing: 1 ml of chloroform-2-propanol (2:1, vol/vol).

Elution: 3 ml of diethyl ether/acetic acid (98:2, vol/vol).

Comments: The ether extract was evaporated to dryness (10 min, 45°C) under a nitrogen stream and the residues was then reconstituted in 70:30 vol/vol methanol–chloroform (3 ml) prior to LC–MS analysis.

3.7.2 Applications of Reversed Phase SPE

Reversed phase (RP) SPE refers to the sorption of organic solutes from a polar mobile phase, such as water or aqueous solvent, into a non-polar stationary phase, such as a C8 or C18 sorbent. The sorption mechanism involves the interaction of the solute within the chains of the stationary phase, i.e. van der Waals or dispersion forces. Some examples of applications of RP SPE are presented in the following.

3.7.2.1 Extraction of Chloroform in Drinking Water [8]

Chloroform or trichloromethane is a byproduct of the chlorination of drinking water. There is no definitive information that chloroform causes cancer in humans. However, the USEPA has listed chloroform as a probable human carcinogen based on evidence that it causes cancer in *in vitro* studies.

Sample: Drinking water.

Compound: Chloroform.

Sorbent: C18 cartridge.

Wetting/Conditioning: 2 ml of acetonitrile followed by 2 ml of distilled water.

Loading: 1 l of a water sample was passed through the cartridges, at a flow rate of 15 ml min^{-1}, by use of a constant flow of dry nitrogen.

Rinsing: None

Elution: 5 ml of pentane, at a flow rate of 2 ml min^{-1}.

Comments: The obtained extracts were dried over sodium sulfate prior to analysis by GC–MS.

3.7.2.2 Pre-concentration of Isopropyl-9H-thioxanthen-9-one (ITX) in Beverages [9]

Isopropyl-9*H*-thioxanthen-9-one (ITX) (Figure 3.6) is used as a photo-inhibitor in UV-cured inks on printed packages of beverages; hence, it may come in contact with the liquid filled in the package. The SPE method was used for sample pre-concentration for a range of samples including milk, juice, tea and yoghurt drinks prior to analysis by LC–tandem mass spectrometry.

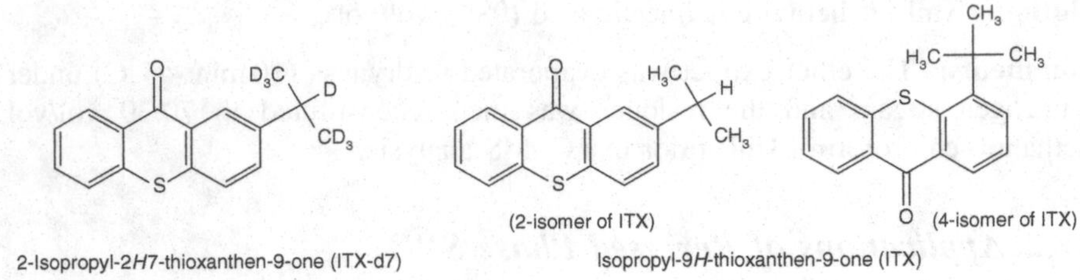

Figure 3.6 Structures of ITX-d7 and ITX (2- and 4-isomers) [9]. Reprinted from *J. Chromatogr., A*, **1143**(1/2), Sun *et al.*, 'Determination of isopropyl-9*H*-thioxanthen-9-one in packaged beverages by solid-phase extraction clean-up and liquid chromatography with tandem mass spectrometry detection', 162–167, Copyright (2007) with permission from Elsevier.

Sample: 10 g of the sample was weighed into the vessel and 100 ml of acetonitrile/water (60:40, vol/vol) containing 1% (vol/vol) of potassium hexacyanoferrate(II) trihydrate and 1% (vol/vol) of zinc acetate was transferred to the sample. The mixture was shaken for 20 min and centrifuged at 4000 rpm for 15 min. 10 ml of the supernatant was removed and diluted to 30 ml with deionized water.

Compounds: ITX-d7 and ITX (2- and 4-isomers).

Sorbent: *m*-Divinylbenzene and *N*-vinylpyrrolidone copolymer, Oasis HLB cartridge.

Wetting/Conditioning: 3 ml of methanol followed by 3 ml of water.

Loading: 6 ml of the diluted sample was loaded onto the cartridge.

Rinsing: 3 ml of water followed by 3 ml of acetonitrile/water (20:80, vol/vol).

Elution: 4 ml of acetonitrile.

Comments: The extract was dried using N_2, reconstituted with 1 ml of acetonitrile/0.1% formic acid (95:5, vol/vol) and then filtered through a 0.45 µm filter paper prior to analysis.

3.7.2.3 Extraction of Pesticides in Washing Water from Olive Oil Processing [10]

The washing step of olive fruits prior to olive oil extraction is carried out in order to remove residual matter, including pesticides. The washing waters from olive oil processing contain a high level of suspension matter and significant amounts of olive oil resulting in a complex matrix to be extracted. The SPE method was developed to separate pesticides from the water matrix followed by GC analysis using thermionic specific detection (TSD) and electron capture detection (ECD).

Sample: Washing waters from olive oil processing filtered under vacuum through filter papers with a pore size of 20 and 8 μm, respectively, followed by a 0.45 μm filter.

Compounds: 28 Organochlorine, organophosphorus and organonitrogen pesticides.

Sorbent: C18 cartridge.

Wetting/Conditioning: 2×5 ml of dichloromethane, 2×5 ml of methanol and 2×5 ml of 'Milli-Q' water.

Loading: 100 μl of a 1 μg ml^{-1} triphenylphosphate (TPP) standard was added to 1 l of a water sample and this solution was slowly passed through the cartridge at a rate ranging from 12–15 ml min^{-1}.

Rinsing: The cartridge was dried by passing air for 15 min and N_2 for another 15 min after sample loading.

Elution: 4×1 ml of dichloromethane for 1 min by gravity and under vacuum for the final elution.

Comments: The extract was filtered over anhydrous Na_2SO_4 followed by washing with dichloromethane and evaporated to dryness, the residue was dissolved by adding 100 μl of a 1 μg ml^{-1} quintozene solution for ECD and 200 μl of a 1 μg mL^{-1} caffeine solution for TSD, and the solution was made up to 1 ml with dichloromethane for analysis.

3.7.3 Applications of Ion Exchange SPE

Ion exchange SPE has been used in the separation of ionic compounds from either a polar or non-polar solvent to the oppositely charged ion exchange sorbent, such as benzenesulfonic acid, propanesulfonic acid and quaternary amines. The separation mechanism involves ionic interaction; hence, a polar compound may be effectively separated from polar solvents, including water, as well as less polar organic solvents.

3.7.3.1 Isolation of Amino Acids from Liquid Samples [11]

Amino acids are the basic constituents of proteins in living organisms. It is necessary to have reliable sample preparation procedures for their isolation from aqueous matrices due to the importance of amino acids in proteins, nutrition, taste and food authentication [11]. SPE procedures employing different types of ion exchangers have been developed as a suitable working procedure for pre-concentration of amino acids from water samples.

Sample: Water samples.

Compounds: Amino acids (some of their structures are shown in Figure 3.7).

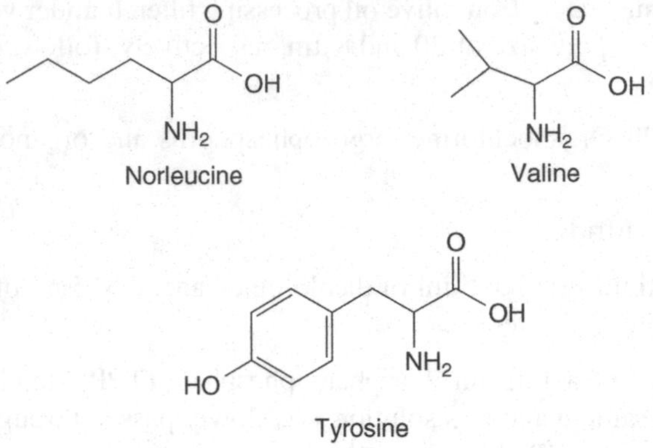

Figure 3.7 Structures of norleucine, valine and tyrosine [11]. Reprinted from *J. Chromatogr., A*, **1150**(1/2), Spanik *et al.*, 'On the use of solid phase ion exchangers for isolation of amino acids from liquid samples and their enantioselective gas chromatographic analysis', 145–154, Copyright (2007) with permission from Elsevier.

Sorbent: Three types of SPE cartridges, consisting of strong anion exchange (SAX–SPE, quaternary amine groups attached to polymeric support/3 ml OASIS MAX, 60/500 mg), weak cation exchange (WCX–SPE, carboxylic groups attached to polymeric support/3 ml BAKERBOND, 60/500 mg) and strong cation exchange (SCX–SPE, sulfonic groups attached to polymeric support/3 mL BAKERBOND, 500 mg).

Wetting/Conditioning: 3 ml of methanol followed by 3 ml of deionized water.

Loading: 2 × 5 ml of a water sample loaded at a flow rate of 1 ml min^{-1}.

Rinsing: None.

Elution: 2.5 ml of 1 M HCl (for SAX–SPE); 1.5 ml of 3 M NH$_4$OH (for WCX–SPE); 2.5 ml of 3 M NH$_4$OH (for SCX–SPE).

Comments: The extracts were analysed by GC–FID. The extraction of amino acids as anions was not successful, and SCX–SPE was found most suitable for isolation of amino acids from water samples.

3.7.3.2 Extraction of Alkylphenols from Produced Water from Offshore Oil Installations [12]

Alkylphenols are commonly found in produced water discharged from offshore oil installations into the sea. Many of them are toxic and able to enter cells of living organisms in the aquatic systems. An SPE anion exchanger was employed in sample preparation for extraction of alkylphenols, followed by GC–MS analysis of their pentafluorobenzoate derivatives.

Sample: Produced water released from offshore oil installations.

Compounds: 14 Alkylphenols.

Sorbent: 6 ml, 150 mg Oasis MAX containing quaternary amine groups.

Wetting/Conditioning: 6 ml of 1:9 vol/vol methanol and *tert*-butyl methyl ether (MTBE) under vacuum, followed by 6 ml of distilled water.

Loading: 100 ml of filtered water samples loaded at a flow rate of 10 ml min^{-1}.

Rinsing: 10 ml of 30% KOH.

Elution: 15 ml of 5% formic acid in methanol.

Comments: The extract was evaporated under a N_2 flow at 39°C to a sample volume of ca. 1 ml, derivatized, diluted 100 times and analysed by GC–MS.

3.7.3.3 Speciation of Cationic Selenium Compounds Present in Leaf Extracts [13]

Selenium can be transported and localized in plants. It is known that the range between selenium as a nutrient and toxicant is very narrow. Hence, it is important to know both total selenium amounts and various selenium species present in plants. This study investigated the presence of two immediate precursors of volatile dimethylselenide in the leaves of *Breassica juncea* by SCX–HPLC–ICP–MS analysis.

Sample: *Brassica juncea* leaf extract.

Compounds: Methylselenomethionine (MeSeMet) and dimethylselenoniumpro-prionate (DMSeP).

Sorbent: 3 ml, 200 mg Strata SCX performed using a 12-port vacuum manifold.

Wetting/Conditioning: 8 bed volumes of methanol followed by 8 bed volumes of 0.75 mM pyridinium formate.

Loading: 1 ml of sample introduced on SCX–SPE and allowed to completely dry.

Rinsing: None.

Elution: 15 bed volumes of 8.0 mM pyridinium formate.

Comments: The effluent was evaporated under a stream of N_2 and then stored at −21°C until analysis by SCX–HPLC–ICP–MS.

3.7.4 Applications of Molecularly Imprinted Polymers (MIPs)

MIPs have been exploited for pre-treatment or removing matrix interferences of samples prior to determination by chromatographic techniques. Development of the sample clean-up technique is aimed for increasing sample throughput, saving cost, simplicity and coupling to both liquid and gas chromatography. Selected applications of MIPs will now be presented.

3.7.4.1 Trace Analysis of Chloramphenicol using MIPs with LC–MS/MS Detection [3]

The use of antibiotic drugs in food-producing animals may cause drug residues in food and result in growing concerns over food safety. Chloramphenicol (CAP) is an antibiotic drug and banned, due to its toxicity, in food-producing animals within the EU and USA. It has potentially fatal side effects (aplastic anemia in humans) and is also suspected of carcinogenity. In this work, MIPs have been developed for pre-concentration of CAP residues prior to detection by LC–MS/MS. The method was applied for identifying CAP in various samples including honey, milk, urine and plasma at below a detection limit of 0.3 μg/kg required by regulatory agencies.

In this example study [3], the MIPs were synthesized using an analogue of CAP as a template molecule in order to eliminate the risk of residual template leaching or bleeding. The MIP SPE method was used to compare the cleanliness of elutes from honey extracts for the different clean-up methods, including a hydrophilic polymer SPE cartridge, 'SupelMIP' SPE chloramphenicol cartridges and LLE. By comparing total ion scans which show all interferences it was clear that 'SupelMIP' SPE chloramphenicol cartridges gave superior sample clean-up (Figure 3.8 (a,b)). It was indicated that the improved cleanliness of the extracts was due to the selective washing solvents used in the SPE sample clean-up. It was also evident that the critical stage in any MIP-based SPE protocol is the selection of appropriate washing solvents, since they allow the high selectivity of the imprinted sites to be revealed. In addition, the method provides more accurate and more sensitive data compared to the other extraction techniques. The procedure is also validated for honey and urine sample matrices according to the European Union (EU) criteria for the analysis of veterinary drug residues.

Pre-treatment for honey samples   1 g of honey and 1 ml of water were combined to get a honey solution. The solution was heated in a water bath at 45°C for 5 min, followed by fortifying with a concentration of 1 μg/l CAP-d5. The solution was transferred to a clean tube and evaporated at 50°C to dryness. The residue was reconstituted in 1 ml of methanol and diluted with 20 ml of water.

Pre-treatment for urine samples   The samples were adjusted with acetic acid to a final pH between 7.0 and 7.5. The samples were then fortified with 1 μg/l CAP-d5. 1 ml of each urine sample was then cleaned up as described for the honey samples. Elution was achieved by applying 2 × 1 ml methanol.

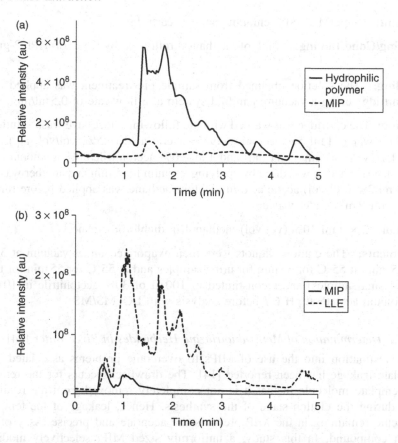

Figure 3.8 (a) Comparison of honey extracts from SupelMIP SPE chloramphenicol and a hydrophilic polymer SPE clean-up. A total ion scan was performed over 100–400 amu. (b) Comparison of honey extracts from SupelMIP SPE chloramphenicol and an LLE sample clean-up. A total ion scan was performed over 150–500 amu [3]. Reprinted from *J. Chromatogr., A*, **1174**(1/2), Boyd *et al.*, 'Development of an improved method for trace analysis of chloramphenicol using molecularly imprinted polymers', 63–71, Copyright (2007) with permission from Elsevier.

Pre-treatment for milk and plasma samples Raw milk samples (5 ml) were centrifuged at $1100 \times g$ for 15 min. The supernatant was collected for application to the SPE cartridge. For plasma samples and semi-skimmed milk, no pre-treatment was required. The samples were fortified with 1 µg/l CAP-d5. 1 ml samples were treated as described for the honey samples except that elution was carried out by applying 2×1 ml 89% (vol/vol) methanol/1% (vol/vol) acetic acid/10% (vol/vol) water.

Sorbent: 'SupelMIP' SPE chloramphenicol cartridges.

Wetting/Conditioning: 1 ml of methanol followed by 1 ml of HPLC-grade water.

Loading: The solution obtained from sample pre-treatment was applied onto the cartridge using a vacuum manifold system at a flow rate of 0.5 ml/min.

Rinsing: The cartridge was washed with the following successive wash solutions: 2 × 1 ml water, 1 ml 5% acetonitrile/95% acetic acid (0.5%, vol/vol, aq.), 2 × 1 ml 1% (vol/vol) ammonia (aq) and 1 ml 20% acetonitrile/80% ammonia (1%, vol/vol, aq). Then, it was dried by applying vacuum for 5 min and another wash of 2 × 1 ml 2% (vol/vol) acetic acid in dichloromethane was applied before further drying for 2 min under vacuum.

Elution: 2 × 1 ml 10% (vol/vol) methanol in dichloromethane.

Comments: The elution aliquots were then evaporated under vacuum at 35°C for 35 min (at 55°C for 35 min for urine samples and at 55°C for 55 min for milk and plasma samples) and reconstituted in 100 μl of 30% acetonitrile in 10 mM ammonium acetate at pH 6.7 before analysis with LC–MS/MS.

3.7.4.2 Determination of Methylthiotriazine Herbicides in River Water [14]

An investigation into the use of MIPs to overcome problems associated with template leakage has been reported [14]. The drawback occurs for the remaining template molecule in that it is not completely removed from the resulting MIP during the elution stage of the synthesis. Hence, leakage of the template molecule remaining in the MIP prevents the accurate and precise assay of the target compound. In this study, a uniformly sized MIP, selectively modified with a hydrophilic external layer (called a restricted access media–molecularly imprinted polymer (RAM-MIP), was prepared for use as a pre-treatment SPE in the simultaneous determination of methylthiotriazine herbicides in river water. The RAM–MIPs were synthesized using a multi-step swelling and polymerization method followed by *in situ* hydrophilic surface modification of the MIPs. A methylthiotriazine skeleton (irgarol) was used as an alternative template molecule, ethylene glycol dimethacrylate as a cross-linker and 2-(trifluoromethyl) acrylic acid (TFMAA) as a functional monomer. The SPE having an RAM-MIP as a sorbent was connected to a column-switching HPLC system, as shown in Figure 3.9. The determination of methylthiotriazine (simetryn, ametryn and prometryn) in river water indicated that the method was accurate and reproducible (Table 3.3). Figure 3.10 shows chromatograms of river water sample spiked and unspiked with methylthiotriazine herbicides. The quantitation limits of simetryn, ametryn and prometryn were 50 pg/ml and the detection limits were 25 pg/ml. The 'recoveries' of simetryn, ametryn and prometryn, at 50 pg/ml were 101%, 95.6% and 95.1%, respectively.

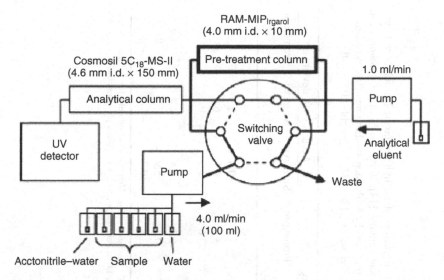

Figure 3.9 The column-switching HPLC system used in this study: solid line, pre-treatment and enrichment step; dashed line, separation step [14]. Reprinted from *J. Chromatogr., A*, **1152**(1/2), Sambe *et al.*, 'Molecularly imprinted polymers for triazine herbicides prepared by multi-step swelling and polymerization method: Their application to the determination of methylthiotriazine herbicides in river water', 130–137, Copyright (2007) with permission from Elsevier.

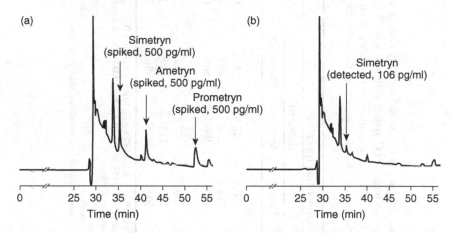

Figure 3.10 Chromatograms of river water sample spiked with methylthiotriazine herbicides (a), and river water sample (b), by a column-switching HPLC system with RAM-MIP as a pretreatment column [14]. Reprinted from *J. Chromatogr., A*, **1152**(1/2), Sambe *et al.*, 'Molecularly imprinted polymers for triazine herbicides prepared by multi-step swelling and polymerization method: Their application to the determination of methylthiotriazine herbicides in river water', 130–137, Copyright (2007) with permission from Elsevier.

Table 3.3 Intra- and inter-day precision and accuracy data for the simultaneous determination of methylthiotriazine herbicides in river water by a column-switching HPLC system with RAM-MIP8Irgarol as a pre-treatment column[a] [14]. Reprinted from *J. Chromatogr., A*, **1152**(1/2), Sambe *et al.*, 'Molecularly imprinted polymers from triazine herbicides prepared by multi-step swelling and polymerization method: Their application to the determination of methylthiotriazine herbicides in river water', 130–137, Copyright (2007) with permission from Elsevier

Added	Concentration (pg/ml) Measured[b]			RSD (%)[c]			Accuracy (% deviation)[d]		
	Simetryn	Ametryn	Prometryn	Simetryn	Ametryn	Prometryn	Simetryn	Ametryn	Prometryn
Intra-day (*n* = 3)									
50	50.4	48.9	51.6	0.8	1.3	3.1	0.8	−2.2	3.1
200	213	212	200	2.7	1.8	3.8	6.7	6.0	−0.0
500	510	496	501	0.8	0.8	2.6	2.1	−0.1	0.2
Inter-day (*n* = 3)									
50	50.7	47.8	52.5	2.0	3.5	6.3	1.4	−4.4	4.9
200	212	215	199	1.4	3.1	0.8	5.9	7.7	−0.6
500	501	495	496	2.3	0.4	2.8	0.2	−1.0	−0.8

[a] Pre-treatment conditions: column, RAM-MIP8$_{Irgarol}$(10 mm × 4.0 mm I.D.); column temperature, 35 °C; injection volume, 100 mL (at 4.0 mL/min for 25 min). Analysis conditions: column, Cosmosil 5C$_{18}$-MS-II(150 mm × 4.6 mm I.D.); column temperature, 35 °C; flow rate, 1.0 mL/min; eluent, 50 mM potassium phosphate buffer–acetonitrile (62:38, v/v, pH 7.0); detection, 230 nm.
[b] Average.
[c] RSD, relative standard deviation.
[d] % deviation = [(concentration measured − concentration added)/concentration added] × 100.

Pre-treatment: The river water samples were stored at $4°C$ and filtered through a 0.45 μm membrane filter.

Sorbent: RAM-MIPs.

Wetting/Conditioning: 'Nanopure' water.

Loading: 100 ml of a river water sample, at a flow rate of 4.0 ml/min.

Elution: The herbicides retained were transferred to an analytical column (Cosmosil 5C18-MS-II packed column) in the back-flush mode using 50 mM potassium phosphate buffer–acetonitrile (62:38, vol/vol, pH 7.0), at a flow rate of 1.0 ml/min.

Comments: The detection was at 230 nm by a UV detector.

3.7.4.3 Extraction of 4-Chlorophenols and 4-Nitrophenol from River Water Samples [15]

The operation of MIP–SPE in an on-line mode coupled to liquid chromatography was investigated [15]. Three different polymers (P1, P2 and P3) were synthesized and evaluated for their potential selectivity for 4-chlorophenols (4-CP) in real water samples. Polymers P1 and P2 were prepared by the 'non-covalent' approach, while polymer P3 was prepared by the 'semi-covalent' approach. In the preparation of P1, 4-CP was used as the template molecule and 4-vinylpyridine (4-VP) the functional monomer. For P2, 4-CP was used as the template molecule and methacrylic acid (MAA) as the functional monomer. For P3, 4-chlorophenyl methacrylate was used as the template molecule and styrene as the additional functional comonomer. Ethylene glycol dimethacrylate (EGDMA) was used as the cross-linker for all polymers. The chromatographic evaluation of the polymers indicated that the 4-VP non-covalent polymer (P1) was the one which showed a clear imprint effect, whereas P2 and P3 did not. In addition, the polymer having 4-CP as a template molecule showed 'cross-reactivity' for 4-chlorophenols and 4-nitrophenol from a mixture containing the 11 priority US EPA (Environmental Protection Agency) phenolic compounds and 4-chlorophenol. The cross-reactivity of the polymer was proved by a washing step with dichloromethane (DCM), as shown in Figure 3.11. The polymer (P1) was then applied for extraction of the river water sample. The results showed that polar phenols cannot be accurately quantified at low levels according to the complex matrix of water-containing humic acids. As can be seen in Figure 3.12, the interference in quantification of the most polar compounds appeared as a broad band at the beginning of the chromatogram. However, the method was modified to use the MIP as a selective sorbent in SPE by including a washing stage with 0.1 ml of DCM (Figure 3.12). This clean-up completely removed the humic band, resulting in the accurate quantification of the compounds selectively retained on the MIP.

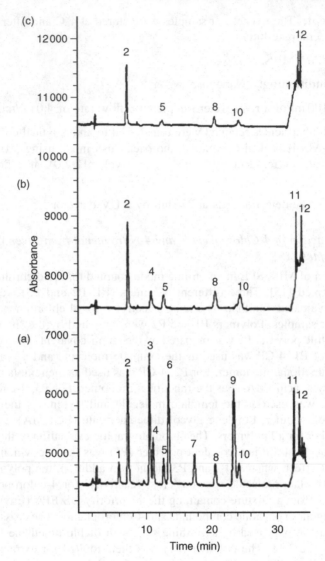

Figure 3.11 Chromatograms obtained by on-line MISPE with the 4-VP non-covalent 4-CP imprinted polymer (P1) of a 10 ml standard solution (pH 2.5) spiked at $10\,\mu g\,l^{-1}$ with each phenolic compound. (a) Without a washing step and (b, c) with a washing step, using 0.1 and 0.3 ml of dichloromethane, respectively: (1) phenol; (2) 4-nitrophenol; (3) 2,4-dinitrophenol; (4) 2-chlorophenol; (5) 4-chlorophenol; (6) 2-nitrophenol; (7) 2,4-dimethylphenol; (8) 4-chloro-3-methylphenol; (9) 2-methyl-4,6-dinitrophenol; (10) 2,4-dichlorophenol; (11) 2,4,6-trichlorophenol; (12) pentachlorophenol [15]. Reprinted from *J. Chromatogr., A*, **995**(1/2), Caro *et al.*, 'On-line solid-phase extraction with molecularly imprinted polymers to selectively extract substituted 4-chlorophenols and 4-nitrophenol from water', 233–238, Copyright (2003) with permission from Elsevier.

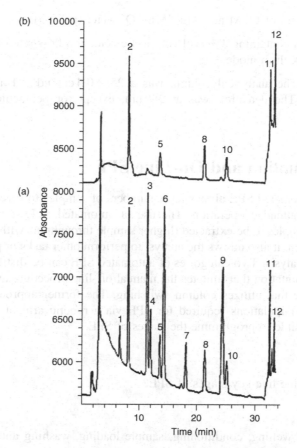

Figure 3.12 Chromatogram obtained by on-line MISPE with the 4-VP non-covalent 4-CP imprinted polymer (P1) of 10 ml of Ebro river water (pH 2.5) spiked at $10 \mu g \, l^{-1}$ with each phenolic compound. (a) Without a washing step and (b) with a washing step using 0.1 ml of dichloromethane. Peak designation as shown in Figure 3.11 [15]. Reprinted from *J. Chromatogr., A*, **995**(1/2), Caro *et al.*, 'On-line solid-phase extraction with molecularly imprinted polymers to selectively extract substituted 4-chlorophenols and 4-nitrophenol from water', 233–238, Copyright (2003) with permission from Elsevier.

Pre-treatment: The river water sample was filtered through 0.45 μm filter, spiked with $10 \mu g \, l^{-1}$ of each compound, and adjusted with HCl to pH 2.5.

Sorbent: MIP (P1).

Wetting/Conditioning: 5 ml acetonitrile (ACN) and 2 ml acidified 'Milli-Q' water with HCl (pH 2.5), at a flow rate of $3 \, ml \, min^{-1}$.

Loading: 10 ml of the spiked water sample was applied to the MIP, at a flow rate of $3 \, ml \, min^{-1}$.

Washing: 0.1 ml of DCM and 4 ml 'Milli-Q' water (pH 2.5).

Elution: ACN containing 1% (vol/vol) acetic acid, at a flow rate of 1 ml min^{-1} and in the back-flush mode.

Comments: The analytical column was a 25×0.4 cm i.d., 'Tracer Extrasil' ODS2, 5 μm. The detection was at 280 nm, except for pentachlorophenol at 302 nm.

3.8 Automation and On-Line SPE

The use of automated SPE allows large numbers of samples to be extracted routinely with unattended operation. The use of automated SPE should therefore allow more samples to be extracted (higher sample throughput) with better precision. In addition, it also allows the analyst to perform other tasks or prepare more samples for analysis. Two categories of automated SPE can be distinguished: the use of instrumentation that imitates the manual off-line procedure and an on-line SPE procedure that utilizes column switching. The former approach 'imitates' the off-line manipulations required for SPE via a robotic arm or autosampler. Thus it is possible to programme the stages of SPE.

DQ 3.5

What are the five key stages of SPE?

Answer

These are wetting, conditioning, sample loading, washing and elution, and then collecting the compound in an appropriate solvent.

The volumes to be used for each stage are programmed into the system as a method. This assumes that the SPE method has been previously well characterized. After completion of this process, the extracted compound is ready for chromatographic analysis.

On-line SPE is the situation where the eluent of the SPE column is automatically directed into the chromatograph (assuming it to be HPLC, although this is not always the case) for separation and quantitation of the compounds of interest. This situation is often described as a 'column switching' or a 'coupled column' technique. The SPE column, or 'pre-column', frequently contains a low-efficiency sorbent which performs a pre-separation of the sample, after which the compound-containing fraction is directed onto a second high-efficiency column for separation and quantitation of the compounds of interest. A simplified diagram for column switching is shown in Figure 3.13. The solvent to wet and pre-condition the sorbent is pumped through the pre-column and then directed to

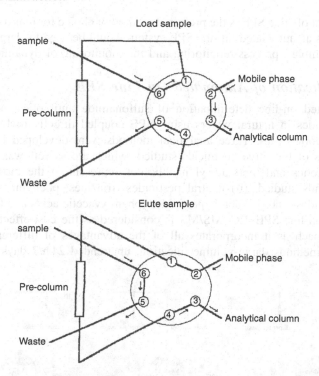

Figure 3.13 Schematic diagram illustrating the principle of column switching. From Dean, J. R., *Extraction Methods for Environmental Analysis*, Copyright 1998. © John Wiley & Sons, Limited. Reproduced with permission.

waste. Then the sample is loaded onto the pre-column and rinsed with an appropriate solvent. In the elution stage, the high-pressure switching valve is rotated so that the mobile phase passes through the pre-column and flushes the compounds onto the analytical separation column. While the analytical separation takes place, the switching valve returns to the 'load' position for re-conditioning of the pre-column ready to start the next sample. Commercial systems are available that utilize this automated on-line procedure.

SAQ 3.5

What are the main advantages to a laboratory of an on-line SPE procedure?

Such advantages (see 'Response to SAQ 3.5') must, of course, be balanced by some disadvantages: the initial time taken to develop a method that is both robust and reliable in terms of both the column technology (pre-column and analytical column) and the equipment used, and the additional capital cost involved. It is envisaged that off-line SPE is the preferred method of choice for non-routine samples, whereas an automated on-line SPE system would be used for large numbers of routine samples, process monitoring and the monitoring of dynamic systems.

3.8.1 Application of Automated On-Line SPE

The automated on-line determination of sulfonamide antibiotics, neutral and acidic pesticides in natural waters using SPE coupled directly to LC–MS/MS has been reported [16]. Three analytical methods were developed for the different groups of bioactive chemicals studied, which are as follows: (i) sulfonamide antibiotics and their acetyl metabolites representing the most polar of the compounds studied, (ii) neutral pesticides (triazines, phenylureas, amides, chloracetanilides) and (iii) acidic pesticides (phenoxyacetic acids and triketones). Automated on-line SPE–LC–MS/MS is considered as the cost-effective instrumental approach as it incorporates all of the advantages of different existing online SPE methods: large-volume injection, unattended 24 h/7 days operation,

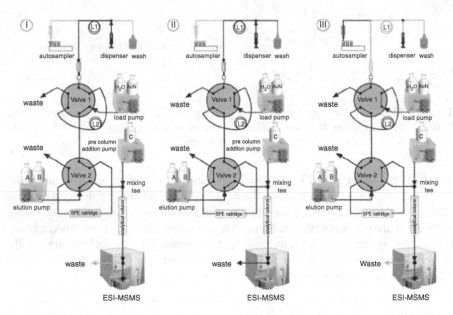

Figure 3.14 Schematic views of the online SPE–LC–MS/MS setup during the three SPE steps: (I) 'loading'; (II) 'enrichment'; (III) 'elution', according to Table 3.4: L1, dispenser loop; L2, sample loop: H₂O, HPLC-grade water: ACN, HPLC-grade acetonitrile: composition of eluents A, B and C, see Table 3.5 [16]. Reprinted from *J. Chromatogr., A*, **1097**(1/2), Stoob *et al.*, 'Fully automated online solid phase extraction coupled directly to liquid chromatography–tandem mass spectrometry: Quantification of sulfonamide antibiotics, neutral and acidic pesticides at low concentrations in surface waters', 138–147, Copyright (2005) with permission from Elsevier.

Table 3.4 Actions of the different components during the SPE steps [16]. Reprinted from *J. Chromatogr., A*, **1097**(1/2), Stoob *et al.*, 'Fully automated online solid phase extraction coupled directly to liquid chromatography–tandem mass spectrometry: Quantification of sulfonamide antibiotics, neutral and acidic pesticides at low concentrations in surface waters', 138–147, Copyright (2005) with permission from Elsevier

SPE step		Time (min)	Valve 1	Valve 2	Dispenser	Load pump	Elution + pre-column addition pump
III	SPE-elution sample n	0		Switch			
		0.5–3.5				Wash sample loop with H_2O	LC-gradient elution sample n
		3.5–5.5				Wash sample loop with AcN	
		5.5–10.5			Buffer addition	Wash sample loop with H_2O	
I	Loading sample n + 1	10.5	Switch	Switch	Charge dispenser and sample loop with sample n + 1		
		10.5–15				Wash SPE cartridge with AcN	LC-gradient elution sample n (continued)
		15–22.5				Conditioning SPE with H_2O	
II	Enrichment sample n + 1	22.5	Switch				
		22.5–33			Wash diluter system	Extract sample n + 1	LC-gradient elution sample n (continued)

Note: The three SPE steps are arranged according to the chromatographic time schedule. During SPE-elution and LC-gradient elution of a given sample n, the next sample n + 1 is loaded and extracted.

Table 3.5 Gradients for the three different methods, where all flow rates are in µl/min [16]. Reprinted from *J. Chromatogr., A*, 1097(1/2), Stoob *et al.*, 'Fully automated online solid phase extraction coupled directly to liquid chromatography–tandem mass spectrometry: Quantification of sulfonamide antibiotics, neutral and acidic pesticides at low concentrations in surface waters', 138–147, Copyright (2005) with permission from Elsevier

Time	Sulfonamides[a]				Neutral pesticides[b]				Acidic pesticides[c]			
	%A	%B	%C	Total flow	%A	%B	%C	Total flow	%A	%B	%C	Total flow
0	5	5	90	400	0	40	60	200	0	40	60	150
4	5	5	90	400								
4.1	20	20	60	250								
20	40	40	20	250	0	90	10	200	0	70	30	150
22	0	80	20	250	0	90	10	200				
23									0	90	10	150
24	0	80	20	250	0	40	60	200				
25									0	90	10	150
26	20	20	60	250					0	40	60	150
28	5	5	90	400								
33	5	5	90	400	0	40	60	200	0	40	60	150

[a] A: water, 20 mM formic acid, pH 2.7; B: methanol; C: water, 10 mM ammonia acetate, pH
[b] A: not used; B: methanol, 20 mM formic acid; C: water, 20 mM formic acid, pH 2.7.
[c] A: not used; B: methanol, 120 mM formic acid; C: water, 120 mM formic acid, pH 2.3.

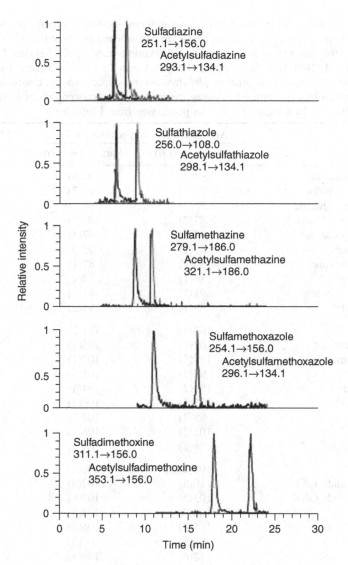

Figure 3.15 Illustrative online SPE–LC–MS/MS chromatogram of a 10 ng/l standard for the sulfonamides and their acetylmetabolites [16]. Reprinted from *J. Chromatogr., A*, **1097**(1/2), Stoob *et al.*, 'Fully automated online solid phase extraction coupled directly to liquid chromatography–tandem mass spectrometry: Quantification of sulfonamide antibiotics, neutral and acidic pesticides at low concentrations in surface waters', 138–147, Copyright (2005) with permission from Elsevier.

Table 3.6 Validation parameters for the three different methods: absolute extraction recovery (%) in nanopure and surface water (in parentheses: combined relative standard uncertainty (%)) and LODs in an environmental sample matrix [16]. Reprinted from *J. Chromatogr., A*, **1097**(1/2), Stoob *et al.*, 'Fully automated online solid phase extraction coupled directly to liquid chromatography–tandem mass spectrometry: Quantification of sulfonamide antibiotics, neutral and acidic pesticides at low concentrations in surface waters', 138–147, Copyright (2005) with permission from Elsevier

Substance	Absolute extraction recovery (%)		LOD (ng/l)
	Nanopure ($n = 6$)	Surface ($n = 6$)	
Acetylsulfadiazine	94(2)	104(3)	5
Acetylsulfadimethoxine	85(1)	92(2)	5
Acetylsulfamethazine	96(1)	95(2)	5
Acetylsulfamethoxazole[a]	87(2)	91(2)	5
Acetylsulfathiazole[a]	95(3)	97(3)	5
Sulfadiazine[a]	87(2)	92(2)	1
Sulfadimethoxine[a]	85(1)	87(1)	1
Sulfamethazine[a]	86(1)	93(1)	1
Sulfamethoxazole[a]	91(1)	87(1)	3
Sulfathiazole[a]	89(1)	91(2)	1
Atrazine[a]	103(1)	111(2)	0.5
Desethylatrazine[a]	101(2)	105(1)	0.5
Dimethenamide[a]	101(3)	107(1)	0.5
Diuron[a]	97(2)	101(1)	0.5
Isoproturon[a]	100(2)	104(1)	0.5
Metolachlor[a]	95(1)	106(1)	0.5
Simazine[a]	99(3)	104(1)	0.5
Tebutam[a]	102(2)	106(1)	0.5
Terbuthylazine[a]	96(2)	104(1)	0.5
2,4-D[a]	106(2)	108(2)	1
Dimethenamide ESA	110(5)	102(3)	3
Dimethenamide OXA	103(5)	103(6)	3
MCPA[a]	102(3)	103(3)	1
Mecoprop[a]	105(3)	106(4)	1
Mesotrione[a]	99(3)	105(5)	2
Metolachlor ESA	112(6)	100(3)	3
Metolachlor OXA	107(5)	107(4%)	3
Sulcotrione[a]	102(3)	104(4)	2

Note: Matrices for extraction recoveries in surface water are creek water for the sulfonamides and lake water for the pesticides.
[a] Isotope-labelled internal standards were used.

low risk for contamination, parallel extraction and separation for high sample throughput, as well as being applicable for very polar compounds. The coupling of the on-line SPE–LC–MS/MS system, using column-switching techniques is shown in Figure 3.14, while the procedure of the on-line SPE process, consisting of three main stages (loading, enrichment and elution) and the gradient, including the composition of the eluents for the three methods is summarized in Table 3.4 and 3.5, respectively. The sample pre-treatment was carried out by filtering with a 250 ml 'bottle-top' filtration unit, using a 0.45 µm cellulose nitrate membrane filter; after that, the sample was adjusted to pH 4 by adding 80 µl of 5 M acetate buffer via the autosampler. The 18 ml sample loop (L2 in Figure 3.14) was loaded with 2 × 9.5 ml samples. Then, the sample enrichment was carried out on an 'Oasis' hydrophilic–lipophilic balance (HLB) extraction cartridge, 20 mm × 2.1 mm i.d., 25 µm particle size using two 6-port valves, with a flow rate of 2 ml min^{-1}. Elution was achieved in the 'back-flush' mode. Consequently, the SPE eluate was mixed with buffered water from the pre-column addition pump prior to the analytical column. A 'Nucleodur' C18 gravity, 125 mm × 2 mm i.d., 5 µm was employed for determination of the sulfonamides and the neutral pesticides, whereas a 'GromSil' ODS 3 CP, 125 mm × 2 mm i.d., 3 µm was used for the acidic pesticides. An illustrative chromatogram of the sulfonamides and their acetyl metabolites is shown in Figure 3.15. To avoid cross-contamination in routine analysis of samples using the same equipment, several cleaning routines were required as follow: (i) washing of the dispenser syringe and loop with a mixture of water and methanol (90/10, vol/vol), (ii) washing of the cartridge with organic solvent and (iii) washing of the analytical column with high-organic-solvent content. The cleanings were implemented after every extraction to remove any residues of the sample, allowing more than 500 samples to be analysed with one extraction cartridge. This enabled a reduction in the extraction cost by more than 75% compared to off-line SPE where SPE cartridges are for single use only. The extraction recovery results indicated that the methods were validated for extraction of the compounds investigated: sulfonamides (85–104%), neutral pesticides (95–111%) and acidic pesticides (99–112%) (see Table 3.6). The limits of detection for the compounds in environmental waters were between 0.5 and 5 ng/l.

SAQ 3.6

It is an important transferable skill to be able to search scientific material of importance to your studies/research. Using your University's Library search engine, search the following databases for information relating to the extraction techniques described in this chapter and specifically the use of solid phase extraction. Remember that often these databases are 'password- protected' and require authorization to access. Possible databases include the following:

- Science Direct;

(continued overleaf)

(continued)

• Web of Knowledge;

• The Royal Society of Chemistry.

(While the use of 'google' will locate some useful information please use the above databases.)

Summary

This chapter describes one of the most important extraction techniques for recovering organic compounds from aqueous samples, i.e. solid phase extraction. The variables in selecting the most effective approach for solid phase extraction are described. Recent developments in new sorbents, e.g. molecularly imprinted polymers, are highlighted and described. The use of solid phase extraction in both off-line and on-line applications is reviewed.

References

1. Moors, M., Massart, D. L. and McDowall, R. D., *Pure Appl. Chem.*, **66**, 277–304 (1994).
2. Somsen, G. W. and de Jong, G. J, Multidimensional chromatography: biomedical and pharmaceutical applications, in *Multidimensional Chromatography*, L. Mondello, A. C. Lewis and K. D. Bartle (Eds), John Wiley & Sons Ltd, Chichester, UK, 2002, pp. 251–307.
3. Boyd, B., Björk, H., Billing, J., Shimelis, O., Axelsson, S., Leonora, M. and Yilmaz, E., *J. Chromatogr. A*, **1174**, 63–71 (2007).
4. Pichon, V., *J. Chromatogr. A*, **1152**, 41–53 (2007).
5. Hong, J., Kim, H.-Y., Kim, D.-G., Seo, J. and Kim, K.-J., *J. Chromatogr. A*, **1038**, 27–35 (2004).
6. Fiorentino, G., Spaccini, R. and Piccolo, A., *Talanta*, **68**, 1135–1142 (2006).
7. Lacaze, J.-P., Stobo, L. A., Turrell, E. A. and Quilliam, M. A., *J. Chromatogr. A*, **1145**, 51–57 (2007).
8. Di Gioia, M. L., Leggio, A., Le Pera, Liguori, A., Napoli, A. and Siciliano, C., *Chromatographia*, **60**, 319–322 (2004).
9. Sun, C., Chan, S. H., Lu, D., Lee, H. M. W. and Bloodworth, B. C., *J. Chromatogr. A*, **1143**, 162–167 (2007).
10. Rubio, M. G., Medina, A. R., Reguera, M. I. P. and de Córdova, M. L. F., *Microchem. J.*, **85**, 257–264 (2007).
11. Spanik, I., Horvathova, G., Janacova, A. and Krupcik, J., *J. Chromatogr. A*, **1150**, 145–154 (2007).
12. Boitsov, S., Meier, S., Klungsoyr, J. and Svardal, A., *J. Chromatogr. A*, **1159**, 131–141 (2007).
13. Yathavakilla, Shah, M., Mounicou, S. and Caruso, J. A., *J. Chromatogr. A*, **1100**, 153–159 (2005).
14. Sambe, H., Hoshina, K. and Haginaka, J., *J. Chromatogr. A*, **1152**, 130–137 (2007).
15. Caro, E., Marce, R. M., Cormack, P. A. G., Sherrington, D. C. and Borrull, F., *J. Chromatogr. A*, **995**, 233–238 (2003).
16. Stoob, K., Singer, H. P., Goetz, C. W., Ruff, M. and Mueller, S. W., *J. Chromatogr. A*, **1097**, 138–147 (2005).

Chapter 4

Solid Phase Microextraction

Learning Objectives

- To be aware of approaches for performing solid phase microextraction of organic compounds from aqueous samples.
- To understand the theoretical basis for solid phase microextraction.
- To understand the practical aspects of solid phase microextraction.
- To appreciate the different methods of operation of solid phase microextraction when used with chromatography.
- To appreciate the different modes of operation of solid phase microextraction.
- To be aware of approaches for performing solid phase microextraction of organic compounds from solid samples.
- To be aware of the practical applications of solid phase microextraction.
- To be aware of the potential of automated solid phase microextraction.

4.1 Introduction

Solid phase microextraction (SPME) is the process whereby an organic compound is adsorbed onto the surface of a coated-silica fibre as a method of pre-concentration. This is followed by desorption of the organic compounds into a suitable instrument for separation and quantitation. The most important stage of this two-stage process is the adsorption of a compound onto a suitably coated-silica fibre or stationary phase. The choice of sorbent is essential, in that it must have a strong affinity for the target organic compounds, so that

Extraction Techniques in Analytical Sciences   John R. Dean
© 2009 John Wiley & Sons, Ltd

Figure 4.1 Solid phase microextraction device.

pre-concentration can occur from either dilute aqueous samples or the gas phase. The range and choice of media available for sorption is ever increasing. Probably the most reported stationary phase for SPME is polydimethylsiloxane (PDMS). This non-polar phase has been utilized for the extraction of a range of non-polar compounds, e.g. benzene, toluene and xylenes (BTEX) from water and air [1]. The fused-silica polydimethylsiloxane-coated fibre is stable at high temperatures. This stability and its small physical diameter and cylindrical geometry allow the fibre to be incorporated into a syringe-like holder (Figure 4.1).

SAQ 4.1

What are the two functions of the SPME holder?

As the normal method of introduction of samples into a gas chromatograph is via a syringe, the use of a syringe-type device for SPME offers no additional complexity. SPME has been exploited most effectively when coupled to gas chromatography (GC), although it has been used for high performance liquid chromatography (HPLC). In the former case, desorption occurs in the hot injector of the gas chromatograph.

SAQ 4.2

How might desorption from the SPME fibre occur in HPLC?

The initial description of SPME will focus on its introduction into the gas chromatograph, as this has been the area initially investigated, and therefore offers the most expansive applications. As we will see later, additional criteria are required when SPME is interfaced to HPLC. The selective nature of the stationary phase of the SPME fibre precludes the introduction of solvent into the gas chromatograph. In addition, no instrument modification is required for GC in terms of, for example, a thermal desorption unit. The heat for desorption from the fibre is provided by the injector of the gas chromatograph.

In the 'unoperable mode', the fused-silica-coated fibre is retracted within the needle of the SPME holder for protection. In operation, however, the coated-silica fibre is exposed to the sample in its matrix. If the sample is aqueous then based full immersion of the coated-silica fibre is required. The active length of the fibre is typically 1 cm. However, it is also possible to extract compounds from the gas phase, e.g. an organic solvent atmosphere in a sealed container (headspace) or the atmosphere in the workplace. In either case, the SPME fibre is exposed to the compound in its matrix (liquid or gaseous) for a pre-selected time period. After sampling, the fibre is retracted within its holder for protection until inserted in the hot injector of the chromatograph. Once located in the hot injector, the fibre is exposed for a particular time to allow for effective desorption of the compounds.

DQ 4.1

How long might desorption take in the injection port of the gas chromatograph?

Answer

This will depend on the volatilities of the organic compounds and their affinities for the SPME fibre coating; however, as the injection port is typically operating at 230°C, desorption will occur rapidly. Usually a period of 2 min is allowed.

As the coating on the fibre is selective towards the compound, it is common to find that no solvent peaks are present in the subsequent GC trace. Unless precautions are made it is important that the delay between the sorption step and the subsequent desorption and analysis step is small. This is because the silica-coated fibre can equally concentrate compounds from the workplace atmosphere (this might be the sample) as it can from the sample or that losses can occur from the fibre. In the first case the risk of contamination from the workplace environment is high. One way to minimize the risk of contamination for aqueous samples at least is to operate SPME using a modified autosampler on the gas chromatograph. In this case, the sealed vials in the autosampler contain the aqueous samples. In operation, the SPME needle can then pierce an individual vial and carry out the sorption stage. This can be immediately followed by insertion into the hot injector of the chromatograph. If an automated system is not available, contamination from the atmosphere can only be eradicated by minimizing the time between extraction and analysis and/or working in a clean room environment. Losses of compound from the SPME fibre can be achieved by employing some form of preservation.

DQ 4.2

How might preservation of organic compounds on the SPME fibre take place?

Answer

Preservation to some extent can occur by cooling the fibre in, for example, a fridge or similar.

4.2 Theoretical Considerations

The partitioning of compounds between an aqueous sample and a stationary phase is the main principle of operation of SPME. A mathematical relationship for the dynamics of the absorption process was developed [2]. In this situation, the amount of compound absorbed by the silica-coated fibre at equilibrium is directly related to its concentration in the sample, as shown below:

$$n = KV_2C_0V_1/KV_2 \times V_1 \tag{4.1}$$

where n is the number of moles of the compound absorbed by the stationary phase, K is the partition coefficient of a compound between the stationary phase and the aqueous phase, C_0 is the initial concentration of compound in the aqueous phase, V_1 is the volume of the aqueous sample and V_2 is the volume of the stationary phase.

As was stated earlier, the polymeric stationary phases used for SPME have a high affinity for organic molecules and hence the values of K are large. These large values of K lead to good pre-concentration of the target compounds in the aqueous sample and a corresponding high sensitivity in terms of the analysis.

However, it is unlikely that the values of K are large enough for exhaustive extraction of compounds from the sample. Therefore SPME is an equilibrium method, but provided proper calibration strategies are followed can provide quantitiative data.

It has been shown [2] that in the case where V_1 is very large (i.e. $V_1 \gg KV_2$) the amount of compound extracted by the stationary phase could be simplified to:

$$n = KV_2C_0 \qquad (4.2)$$

and hence is not related to the sample volume. This feature can be most effectively exploited in field sampling. In this situation, compounds present in natural waters, e.g. lakes and rivers, can be effectively sampled, pre-concentrated and then transported back to the laboratory for subsequent analysis.

The dynamics of extraction are controlled by the mass transport of the compounds from the sample to the stationary phase of the silica-coated fibre. The dynamics of the absorption process have been mathematically modelled [2]. In this work, it was assumed that the extraction process is diffusion-limited. Therefore, the amount of sample absorbed, plotted as a function of time, can be derived by solving Fick's Second Law of Diffusion (see Chapter 11). A plot of the amount of sample absorbed versus time is termed the extraction profile.

DQ 4.3

How might the dynamics of extraction be increased?

Answer

The dynamics of extraction can be increased by stirring the aqueous sample.

4.3 Experimental

The most common approach for SPME is its use for GC, although as will be seen later its coupling to HPLC has been reported. The SPME device consists of a fused-silica fibre coated with a stationary phase, e.g. polydimethylsiloxane. In addition, other stationary phases are available for SPME (Table 4.1). The small size and cylindrical geometry allow the fibre to be incorporated into a syringe-type device (Figure 4.1). This allows the SPME device to be effectively used in the normal 'un-modified' injector of a gas chromatograph. As can be seen in Figure 4.1, the fused-silica fibre (approximately 1 cm) is connected to a stainless-steel tube for mechanical strength. This assembly is mounted within the syringe barrel for protection when not in use. For SPME, the fibre is withdrawn into the syringe barrel, then inserted into the sample-containing vial for either solution or

Table 4.1 Commonly available SPME fibres [3]

Stationary phase	Thickness (μm)	Description	Comments
Polydimethylsiloxane (PDMS)	100 30 7	Non-bonded Non-bonded Bonded	} High capacity, for volatile and apolar compounds, e.g. VOCs Higher desorption temperatures. For semivolatile and apolar compounds, e.g. PAHs
Polydimethylsiloxane/ divinylbenzene (PDMS/DVB)	65 60 65	Partially crosslinked Partially crosslinked Highly crosslinked	} For many polar compounds, especially amines
Polyacrylate (PA)	85	Partially crosslinked	High capacity. For both polar and apolar compounds, e.g. pesticides and phenols
Carboxen/ polydimethylsiloxane (CAR/PDMS)	75 85	Partially crosslinked Highly crosslinked	} High retention for trace analysis. For gaseous/volatile compounds
Carbowax/ divinylbenzene (CW/DVB)	65 70	Partially crosslinked Highly crosslinked	} Low temperature limit. For polar compounds, especially alcohols
Carbowax/templated resin (CW/TPR)	50	Partially crosslinked	} For HPLC applications, e.g. surfactants
Divinylbenzene/ Carboxen/PDMS (DVB/CAR/PDMS)	50/30	Highly crosslinked	} Ideal for broad range of compound polarities, good for C2–C20 range

air analysis. At this point, the fibre is exposed to the compound(s) by pressing down the plunger, for a pre-specified time.

DQ 4.4

How long might the fibre be exposed in the sampling mode?

Answer

This can vary depending upon the organic compounds to be sampled and their volatilities. However, exposure might be from a few minutes to over 20 min.

After this pre-determined time interval, the fibre is withdrawn back into its protective syringe barrel and withdrawn from the sample vial. The SPME device is then inserted into the hot injector of the chromatograph and the fibre exposed for a pre-specified time.

DQ 4.5

How long might the fibre be exposed in the desorb mode?

Answer

Typically, no more than 2 min in the injection port of the gas chromato-graph.

The heat of the injector desorbs the compound(s) from the fibre prior to GC separation and detection. SPME can be done manually or by an autosampler. As the exposed fibre is an active site for adsorption of not only compounds of interest but also air-borne contaminants, it is essential that the SPME fibre is placed in the hot injector of the gas chromatograph prior to adsorption/desorption of compounds of interest to remove potential interferents.

For HPLC analysis using SPME, a separate interface is required. The actual adsorption of compounds onto the SPME fibre is the same for both GC and HPLC with the difference being the means of desorption. Unlike in GC, no hot injector is available to desorb the compounds from the fibre. For HPLC therefore, desorption is achieved using the mobile phase of the system. In order to achieve this a separate interface is required. The procedure is as follows. Before transferring the fibre into the desorption chamber of the interface, the injection valve is placed in the 'load' position. The fibre is then introduced into the desorption chamber by lowering the syringe plunger. The two-piece PEEK union is then closed tightly. The valve is then switched to the 'injection' position, and the desorption procedure started. Solvents from the HPLC pump pass through the desorption chamber in an 'upstream direction' to avoid air bubbles being introduced to the analytical column and disturbing the detector. Compounds that were absorbed by the fibre are then desorbed by the organic solvent and carried to the separation column. Analytical column separation is then initiated and a solvent programme applied to achieve good analytical separation of the compounds of interest.

4.4 Methods of Analysis: SPME–GC

4.4.1 Direct Immersion SPME: Semi-Volatile Organic Compounds in Water

The application of SPME for analysis of semi-volatile organic compounds (specifically PAHs) in aqueous samples has been reported by several authors [4–8]. In one paper [5] it was possible to demonstrate that 16 PAHs could be simultaneously extracted from aqueous sample using a 100 μm PDMS fibre followed by GC–MS analysis. The following conditions were used [5]: absorption time, 45 min with agitation by ultrasonication; desorption temperature, 220°C at the injector port of the gas chromatograph. The mass spectrometer was operated in the electron impact (EI) mode with an ion source temperature of 250°C.

Figure 4.2 shows the GC–MS chromatograms obtained using both SPME and direct injection of a standard containing 19 PAHs and indicates that peak resolution and response are comparable for most of compounds studied. Linearity of the method was investigated over the range 0.01–10 μg l^{-1}. The limit of detection (LOD) of the SPME technique was between 1 and 29 ng l^{-1}. The precision of the method expressed as % RSD was generally <20%.

4.4.2 Headspace SPME: Volatile Organic Compounds (VOCs) in Water

In addition to placing the SPME fibre directly into the aqueous sample it is possible, provided that the compounds are volatile, to use a headspace approach to SPME. Initial work on headspace SPME was reported [9] in 1993 in which it was reported that the sampling time for BTEX in water can be reduced to 1 min compared to direct SPME sampling of the aqueous phase. At ambient temperatures, the headspace SPME approach can be applied to compounds with Henry's constants above 90 atm cm^3 mol^{-1}, i.e. 'three-ring' PAHs or more volatile species. It was also suggested that the equilibration times for less volatile compounds can be shortened significantly by agitation of both aqueous phase and headspace, reduction of headspace volume and by increasing the temperature. It was also reported [9] that headspace SPME could be carried out above soil or sewage samples for PAHs.

Recently a rapid method for extracting and analysing 27 volatile organic compounds, including disinfection by-products in drinking water using HS–SPME and GC/TOF–MS with a split/splitless injector, has been reported [10]. SPME fibres with different coatings, including polydimethylsiloxane (PDMS) (7 μm and 100 μm), carboxen/polydimethylsiloxane (CAR/PDMS), polydimethylsiloxane/divinylbenzene (PDMS/DVB) and DVB/CAR/PDMS, were utilized. The optimum conditions obtained were as follows: DVB/CAR/PDMS best fibre coating (as shown in Figure 4.3); 1% salt concentration; 2 min extraction time;

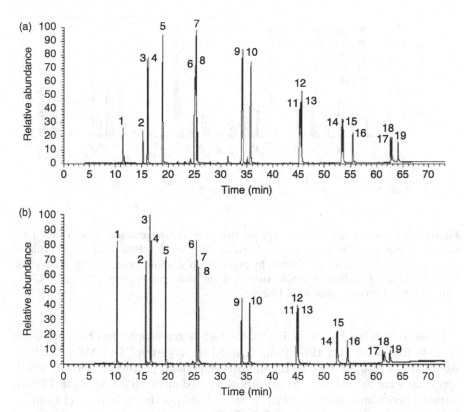

Figure 4.2 GC–MS chromatograms obtained from (a) an SPME extraction from a 1 ml solution of 19 PAHs (10 µg/l) in water and from (b) a 1 µl injection of 19 PAH standards (10 ng/µl) in hexane. Peak numbers correspond to (1) naphthalene, (2) acenaphthylene, (3) acenaphthene-d_{10}, (4) acenaphthene, (5) fluorene, (6) phenanthrene-d_{10}, (7) phenanthrene, (8) anthracene, (9) fluoranthene, (10) pyrene, (11) benz[a]anthracene, (12) chrysene-d_{12}, (13) chrysene, (14) benzo[b]fluoranthene, (15) benzo[k]fluoranthene, (16) benzo[a]pyrene, (17) indeno[1,2,3-cd]pyrene, (18) dibenz[a,h]anthracene and (19) benzo[ghi]perylene [5]. Reprinted from *Anal. Chim. Acta*, **523**(2), King *et al.*, 'Determination of polycyclic aromatic hydrocarbons in water by solid-phase microextraction–gas chromatography–mass spectrometry', 259–267, Copyright (2004) with permission from Elsevier.

35°C extraction temperature; 45 s GC run time for the GC/TOF–MS instrument. It was concluded that the VOCs detection limits were lower than their maximum concentration levels (MCLs) allowed in drinking water and their precisions at 100 ng ml^{-1} were generally good (Table 4.2). In addition, the method developed for analysing VOCs in water samples can be applied as an alternative for the 'purge and trap EPA Method 624'.

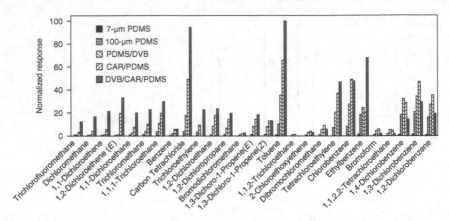

Figure 4.3 Effect of the coating type of the fibre on the extraction of VOCs [10]. Reprinted from *J. Chromatogr*., *A*, **1201**(2), Niri *et al.*, 'Fast analysis of volatile organic compounds and disinfection by-products in drinking water using solid-phase microextraction–gas chromatography/time of flight mass spectrometry', 222–227, Copyright (2008) with permission from Elsevier.

Other work related to extracting VOCs from water samples has been presented [11]. In this work [11] HS–SPME coupled to 'cryo-trap' GC–MS procedures were used to analyse trace BTEX in water. Optimum parameters for this SPME approach were as follows: 75 μm carboxen/polydimethylsiloxane (CAR/PDMS) coated fiber, ionic strength (0.267 g ml^{-1} NaCl), extraction time (15 min, at 25°C), and desorption (2 min, at 290°C). The linearity of the method extended to over five orders of magnitude for all of the compounds. Good analytical performance was obtained, as shown in Table 4.3. A mass ion chromatogram of a ground water sample is shown in Figure 4.4.

4.4.3 Analysis of Compounds from Solid Matrices

The use of SPME to quantify the level of pollutants in soils and sediments has been presented by several authors [12–16]. The intention is that, for direct immersion SPME, a known quantity of soil is stirred with water (or hot water) and then to expose the SPME fibre directly to the resultant slurry prior to analysis. An initial attempt to demonstrate this application was presented in 1995 [17]. Advances in this approach have included use of ultrasonic extraction coupled with SPME for the extraction of two agrochemical fungicides, vinclozolin and dicloran, in soil samples [14]. Two different extraction approaches were compared; water ultrasonic extraction/SPME and acetone ultrasonic extraction/SPME prior to analysis by GC–MS. A soil sample (5 g) mixed with solvent (30 ml water containing 5% vol/vol acetone and 5 ml of acetone for the former and the latter approaches, respectively) was homogenized by sonication for 30 min. The polyacrylate 85 μm

Table 4.2 Analytical performance criteria of the method and maximum concentration level (MCL) for VOCs [10]

Analyte	Precision at 100 ng/ml (%)	Estimated LOD (ng/ml)	MCL (ng/ml)
Trichloromonofluoromethane	10.8	0.477	– [a]
Dichloromethane	13.2	0.278	5
1,1-Dichloroethene	3.3	0.196	7
1,2-Dichloroethene(E)	2.6	0.196	70
1,1-Dichloroethane	4.7	0.112	– [a]
Trichloromethane	11.8	0.078	80
1,1,1-Trichloroethane	3.8	0.071	200
Benzene	2.4	0.066	5
CarbonTetrachloride	4.7	0.044	5
Trichloroethylene	4.8	0.044	5
1,2-Dichloroethane	2.9	0.065	5
1,2-Dichloropropane	3.7	0.025	5
Bromodichloromethane	6.1	0.029	80
1,3-Dichloro-1-propene(E)	6.1	0.029	– [a]
1,3-Dichloro-1-propene(Z)	1.6	0.035	– [a]
Toluene	1.5	0.038	1000
1,1,2-Trichloroethane	1.8	0.038	5
2-Chloroethoxyethene	3.0	0.015	– [a]
Dibromochloromethane	3.8	0.015	80
Tetrachloroethylene	4.4	0.044	5
Chlorobenzene	1.4	0.063	100
Ethylbenzene	3.2	0.022	700
Tribromomethane	1.8	0.049	80
1,1,2,2-Tetrachloroethane	8.3	0.024	– [a]
1,4-Dichlorobenzene	8.8	0.032	75
1,3-Dichlorobenzene	14.4	0.031	75
1,2-Dichlorobenzene	6.7	0.022	600

[a]No MCL has been established for this specific contaminant.

fibre was used for isolation of the fungicides. The optimized SPME conditions were: 45 min sampling time, 5 min desorption time, 960 rpm stirring rate and 25% (wt/vol) NaCl. This demonstrated that acetone extraction/SPME was superior in terms of recovery, precision and limit of detection (Table 4.4). In addition, comparison between the acetone ultrasonic extraction/SPME and classical LLE was made and indicated that the former was less influenced by sample matrix but offered similar performance in terms of recovery (Figure 4.5).

An alternative strategy to solid analysis is to use SPME to extract compounds from the headspace above a sample. The utilization of HS–SPME has been presented by several authors [18–25]. For volatile compounds, headspace SPME is preferred over direct immersion SPME because of its longer lifetime. In the

Table 4.3 Analytical performance criteria obtained using HS–SPME coupled to 'cryo-trap' GC–MS [11]

Compound	Linear range ($\mu g\,l^{-1}$)	r^2	Limit of detection ($ng\,l^{-1}$)	Precision at $0.1\,\mu g\,l^{-1}$ ($\%$, $n = 9$)	Precision at $40\,\mu g\,l^{-1}$ ($\%$, $n = 9$)	Method detection limits as required by the USEPA ($ng\,l^{-1}$)
Benzene	0.0001–50	0.998	0.04	11.2	5.2	30
Toluene	0.0001–50	0.998	0.02	8.9	4.5	80
Ethylbenzene	0.0001–50	0.996	0.05	11.6	6.8	60
m- and *p*-Xylene	0.0001–50	0.999	0.01	8.4	3.1	90
o-Xylene	0.0001–50	0.998	0.02	7.8	4.8	60

Table 4.4 Recovery (%, $n = 3$), LODs and RSDs for determination of fungicides in spiked soils by two different extraction approaches followed by GC–MS [14]. Reprinted from *Anal. Chim. Acta*, **514**(1), Lambropoulou and Albanis, 'Determination of the fungicides vinclozolin and dichloran in soils using ultrasonic extraction coupled with solid-phase microextraction', 125–130, Copyright (2004) with permission from Elsevier

Fungicides	Water extraction/SPME				Acetone extraction/SPME			
	Linear range[a] (r)	Recovery[a] (%) (200 ng g⁻¹)	LOD (ng g⁻¹)	RSD (%)	Linear range[a] (r)	Recovery[a] (%) (200 ng g⁻¹)	LOD (ng g⁻¹)	RSD (%)
Dicloran	0.988	69	13[b]	11.4[d]	0.994	94	3[c]	5.6[d]
Vinclozolin	0.990	64	8[b]	14.2[d]	0.996	91	2[c]	7.4[d]

[a]Linear curves were constructed using five samples between 25 and 500 ng g⁻¹ (25, 50, 100, 250 and 500 ng g⁻¹) and between 10 and 500 (10, 50, 100, 250 and 500 ng g⁻¹) for water/SPME and acetone/SPME methods, respectively.
[b]Calculated from the chromatograph of the sample spiked at 25 ng g⁻¹ concentration level.
[c]Calculated from the chromatograph of the sample spiked at 10 ng g⁻¹ concentration level.
[d]The overall precision was obtained at three concentration levels (25, 100, and 200 ng g⁻¹).

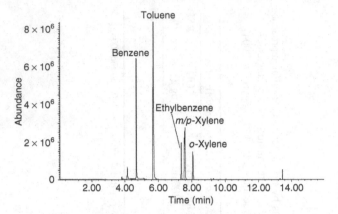

Figure 4.4 A chromatogram of a ground water sample analysed by HS–SPME 'cryo-trap'–GC–MS [11]. Reprinted from *Chemosphere*, **69**(9), Lee *et al.*, 'Determination of benzene, toluene, ethylbenzene, xylenes in water at sub-ng^{-1} levels by solid-phase microextraction coupled to cryo-trap gas chromatography–mass spectrometry', 1381–1387, Copyright (2007) with permission from Elsevier.

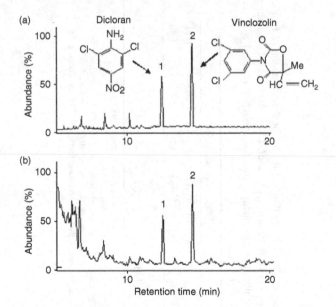

Figure 4.5 GC–SIM–MS chromatograms obtained by (a) the acetone/SPME procedure and (b) liquid–liquid extraction in spiked (250 ng g^{-1}) soil [14]. Reprinted from *Anal. Chim. Acta*, **514**(1), Lambropoulou and Albanis, 'Determination of the fungicides vinclo-zolin and dicloran in soils using ultrasonic extraction coupled with solid-phase microex-traction', 125–130, Copyright (2004) with permission from Elsevier.

case of direct immersion, the fibre coating can be damaged by the complex sample matrix, as the fibre is directly immersed into the sample solution. It is also reported that headspace is more selective than direct immersion [26]. Recently the use of multiple HS–SPME to remove the matrix effect in order to determine BTEX in a contaminated soil and a certified soil has been reported [27]. This approach employed several consecutive extractions from the same sample using HS–SPME coupled to GC–FID. A 75 μm carboxen/polydimethylsiloxane (CAR/PDMS) fibre was used. A soil suspension (15–20 mg soil) in water (600 μl) incubated at 30°C was placed in a 20 ml headspace glass vial, and agitated at 400 rpm for 10 min before extraction. Sampling of BTEX was carried out for 20 min in three consecutive extractions and the desorption time was allowed for 10 min. The HS–SPME–GC–FID chromatograms of the certified soil are shown in Figure 4.6. BTEX concentrations (Table 4.5) were calculated by interpolating

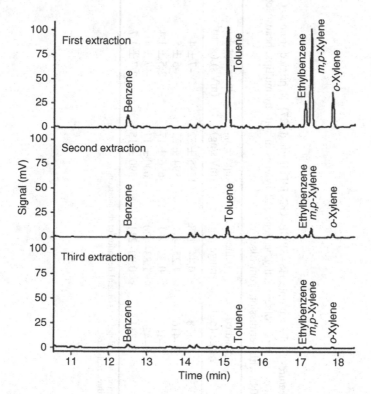

Figure 4.6 Chromatograms of three consecutive HS–SPME extractions of BTEX from a certified soil sample [27]. Reprinted from *J. Chromatogr., A*, **1035**(1), Ezquerro *et al.*, 'Determination of benzene, toluene, ethylbenzene and xylenes in soils by multiple headspace solid-phase microextraction', 17–22, Copyright (2004) with permission from Elsevier.

Table 4.5 Features of the multiple HS–SPME–GC–FID method [27]. Reprinted from *J. Chromatogr.*, A, **1035**(1), Ezquerro *et al.*, 'Determination of benzene, toluene, ethylbenzene and xylenes in soils by multiple headspace solid-phase microextraction', 17–22, Copyright (2004) with permission from Elsevier

Compound	Studied range (ng)	Linear range (ng)	Slope ± s_m (mV s/ng)	Intercept ± s_b (mV s) (× 10^3)	LOD (ng)	R^2	RSD[a] (%) (mass level, ng)
Benzene	0–158	0.44–158	1585 ± 52	−7 ± 4	0.2	0.994	3.9 (66)
Toluene	0–416	1.25–416	894 ± 22	−5 ± 5	1.0	0.996	6.9 (260)
Ethylbenzene	0–161	0.36–161	636 ± 17	−2.5 ± 1.4	0.2	0.996	3.2 (67)
m,p-Xylene	0–420	1.83–420	600 ± 17	−7 ± 4	1.0	0.995	6.2 (260)
o-Xylene	0–211	0.90–211	590 ± 15	−2.7 ± 1.7	0.4	0.996	6.0 (132)

s_m: standard deviation of the slope. s_b: standard deviation of the intercept.
[a]Calculated from three replicates.

the total peak area found for the soils in the calibration graphs obtained from aqueous BTEX solutions. The accuracy of the method was checked by analysing a certified soil and it was found that the concentrations of toluene, ethylbenzene, *o*-xylene and *m,p*-xylenes measured were in good agreement with the certified values.

4.4.4 Other SPME–GC Applications

4.4.4.1 Analysis of Pesticides in Aqueous Samples

The analysis of pesticides has been widely investigated in terms of SPME applications [28–34]. Recently, the limits of quantitation for 18 organochlorines in ground water samples in the range from 4.5×10^{-3} to 1.5 ng l^{-1} with a 50/30 μm DVB–CAR–PDMS fibre coupled with a gas chromatograph equipped with an electron capture dectector and a split/splitless injector were reported [35]. Good precisions were obtained using this approach with typical relative standard deviations (RSDs) ranging from 0.5 to 4.6% for 1.5, 3.0 and 6.0 ng l^{-1} organochlorines in water. The optimized parameters used for SPME were: extraction time (45 min), desorption time (held for 2 min, at 260°C of the GC injector), pH (6.0), ionic strength (no salt addition) and stirring speed (60% of the maximum speed of a magnetic stirrer). The GC detector and injector temperatures were maintained at 300 and 260°C, respectively. The total time for the GC run was 32 min. Figure 4.7 shows a chromatogram of organochlorine pesticides in a ground water sample.

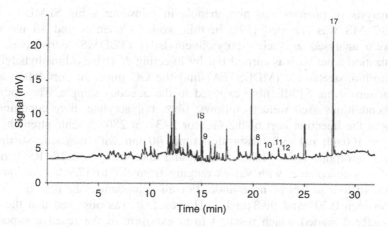

Figure 4.7 Chromatogram obtained by SPME–GC–ECD analysis of a ground water sample: IS, internal standard; (8) endosulfanI; (10) dieldrin; (11) endrin; (12) endosulfanII; (17) endrinketone [35]. Reprinted from *Talanta.*, **72**(5), Júunior and Ré-Poppi, 'Determination of organochlorine pesticides in ground water samples using solid-phase microextraction by gas chromatography–election capture detection', 1833–1841, Copyright (2007) with permission from Elsevier.

4.4.4.2 Determination of Organochlorine Pesticides (OCPs) in Fish Tissue

A clean-up and pre-concentration procedure for organochlorine pesticide determination in fish tissue using SPME followed by GC–ECD has been described [36]. Fish muscle tissue (10 g wet weight) ground with a 4-fold excess of activated anhydrous sodium sulfate was Soxhlet-extracted with 300 ml of a 1:1 vol/vol hexane:acetone solvent mixture for 16 h, in order to remove the fatty matrix, and concentrated under vacuum rotary evaporation to 100 ml prior to the SPME procedure. Aliquots of 1 ml of the organic extract, evaporated to dryness and re-dissolved in 5% vol/vol methanol/water, were then taken for SPME extraction using the following conditions: fibre, 100 μm PDMS; fibre conditioning, heating at the injection port of the gas chromatograph for 1 h at 260°C; time of immersion of fibre in sample, 30 min at ambient temperature (25°C); agitation during extraction, using a stirring bar and a magnetic stirrer; desorption time, 5 min at 260°C. It was found that the LODs obtained from the fish tissue varied from 0.1 to 0.7 ng g^{-1}, recoveries were over 70% for all OCPs (at a concentration level of 10 ng g^{-1}) and the RSDs ranged from 6 to 28%. In addition, the developed method was applied to the analysis of OCPs in CRM 430 (a matrix of pork fat), using the standard-addition method; the measured results were in good agreement with the certified values. Typical chromatograms of the 16 OCPs obtained from the SPME–GC–ECD analysis of a fish tissue organic extract are shown in Figure 4.8.

4.4.4.3 Analysis of Phenols and Nitrophenols in Rainwater

The analysis of phenols and nitrophenols in rainwater using SPME coupled with GC–MS was reported [37]. In this work, 4 phenols and 16 nitrophenols were analysed as their *t*-butyldimethylsilyl (TBDMS) derivatives. The derivatization reaction was carried out by injecting *N*-(*t*-butyldimethylsilyl)-*N*-methyltrifluoroacetamide (MDBSTFA) into the GC injection port followed by introduction of the SPME fibre exposed to the aqueous sample. The optimum SPME conditions used were as follows: fibre, polyacrylate; fibre conditioning, heating at the injection port of the GC for 2–3 h at 280°C; ionic strength, 75 g NaCl per 100 ml; pH, 3.0; absorption time, 40 min with magnetic stirring at 400 rpm; desorption time, 5 min. It was found that precision (as % RSD) of the method was acceptable, with values ranging from 8.7 to 17.9%. The linearity extended to four orders of magnitude. For all compounds, the detection limits were between 0.208 and 99.3 μg l^{-1}. However, it was observed that the fibre was rapidly degraded which resulted from exposure to the reactive vapour of the derivatizing agent.

4.4.4.4 Analysis of Furans in Foods

The feasibility of HS–SPME coupled to GC–ion trap–mass spectrometry (GC–IT–MS) for analysis of furans in different heat-treated carbohydrate-rich

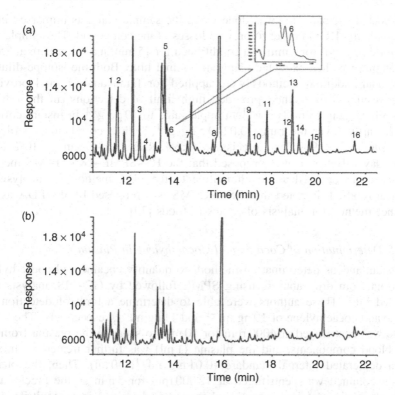

Figure 4.8 Chromatograms obtained by SPME–GC–ECD of: (a) fish tissue extract spiked with OCPs and (b) unspiked fish tissue extract. Peak assignments: (1) HCB; (2) α-HCH; (3) β-HCH; (4) γ-HCH; (5) δ-HCH; (6) heptachlor; (7) aldrin; (8) isodrin; (9) p,p'-DDE; (10) endosulfan α; (11) dieldrin; (12) endrin; (13) p,p'-DDD; (14) endosulfan β; (15) p,p'-DDT; (16) methoxychlor [36]. Reprinted from *J. Chromatogr., A*, **1017**(1/2), Fidalgo-used *et al.*, 'Solid-phase microextraction as a clean-up and preconcentration procedure for organochlorine pesticides determination in fish tissue by gas chromatography with electron capture determination', 35–44, Copyright (2003) with permission from Elsevier.

food samples was proposed [38]. Six commercially available fibres were investigated and it was concluded that a 75 μm carboxen/polydimethylsiloxane coating was the most effective for the extraction of furans. Operating parameters affecting the SPME extraction and desorption process were optimized and include: extraction temperature and time (25°C, 30 min), ionic strength (20% wt/wt NaCl), headspace and aqueous volume ratio (25 ml/15 ml in a 40 ml glass vial), stirring speed (1200 rpm), and desorption temperature and time (275°C, 2 min). The SPME procedure was carried out by placing an optimal amount of the homogenized sample solution in a 40 ml screwed cap glass vial fitted with silicone PTFE-septa containing 4 g of sodium chloride, 15 ml of water and a

PTFE-coated stir bar. This was done when the sample vial was immersed in an ice/water bath ($4°C$) in order to prevent losses of the compound. The sample vial was 'vortex-mixed' for 3 min and conditioned for 15 min in a water bath at $25°C$. The sample was then extracted using an optimal fibre. Both the isotope-dilution and standard-addition methods were applied for furan analysis and provided similar results. This method provided high limit of detections (in the low pg g^{-1} level, ranging from 8 pg g^{-1} in apple juice to 70 pg g^{-1} in instant coffee), good linearity (over the range 0.02–0.5 ng g^{-1}, with a correlation coefficient (r^2) higher than 0.999) and precisions (<6% RSD 'run-to-run', <10% RSD 'day-to-day'). Hence, it was proposed that the HS–SPME–GC–IT–MS method developed can be used as an alternative to the FDA method for analysis of furans in foods. It is noted that HS–GC–MS was proposed by the FDA as the reference method for analysis of furans in foods [39].

4.4.4.5 Determination of Cocaine and Cocaethylene in Plasma

The 'simultaneous determination method' to quantify cocaine and cocaethylene in plasma from drug abusers using SPME followed by GC–MS analysis was proposed [40]. These authors were able to determine a limit of detection for cocaine and cocaethylene of 19 ng ml^{-1} and 11 ng ml^{-1}, respectively. The blood sample was centrifuged at 4000 rpm for 10 min to separate the plasma from the other blood components, and the plasma (1 ml) was further treated by mixing with a deuterated internal standard (0.01 mg ml^{-1}, 40 ml). Then, the plasma solution obtained was centrifuged at 12 000 rpm for 5 min as the precipitation of the plasmatic proteins occurred when it was dissolved in acetonitrile. Four hundred microlitres of clear solution were taken to dissolve with 50 mg sodium chloride and mixed with 200 µl of borax buffer (pH 9). The coating fibre used was 100 µm PDMS as it was previously proved to be suitable for extraction of compounds of medium to low polarity. The authors used a 25 min absorption time and 5 min desorption time at $250°C$ GC injection. The mass spectrometer was run in the selected ion monitoring (SIM) mode. The method showed good linearity (in the range of 25–1000 ng ml^{-1}) and precisions (<15% RSD at all concentrations).

4.4.4.6 Determination of Fluoride in Toothpaste

A rapid method for the determination of fluoride in toothpaste employing HS–SPME, followed by GC–FID, was reported [41]. Trimethylchlorosilane (TMCS) was used as the derivatization reagent to form volatile trimethylfluorosilane (TMFS). The optimization of the SPME procedure was investigated and concluded as follows: 75 µm carboxen/polydimethylsiloxane (CAR/PDMS) coated fibre, absorption time (10 min at $22°C$), stirring speed (500 rpm) and desorption time (4 min at the GC injection port at $200°C$). The linearity of the method was evaluated over the range of 0.25 to 1.25 mg ml^{-1} fluoride showing a

Table 4.6 Comparison between HS–SPME and LLE followed by GC–FID for determination of fluoride in toothpaste [41]

Description	HS-SPME	LLE
Sample weight	800 mg	800 mg
Amount of TMCS	30 µl	2 ml
Amount of solvent	–	5 ml
Time of derivatization reaction	10 min	15 min
Time of extraction	10 min	35 min
Contents of NaF found in toothpaste sample containing 0.321% NaF	0.326% ($n = 3$)	0.324% ($n = 3$)
Linearity (r^2) over the range of 0.25 to 1.25 mg ml^{-1}	0.991	0.992
Precision (as % RSD)	11.94% ($n = 9$)	10.08% ($n = 10$)

correlation coefficient (r^2) of 0.991. The limit of detection was found to be 6 µg ml^{-1} and the precision was good (11.94% RSD, $n = 9$). Comparison between HS–SPME and liquid–liquid extraction (LLE) was made with respect to their linearity, precision and accuracy (Table 4.6). It was found that the two extraction procedures gave very similar results. However, the authors recommended that SPME should be used for routine determination as it has some advantages over LLE, i.e. SPME is inexpensive, fast, simple and eliminates the costs and hazards associated with the use of large amount of organic solvents.

4.5 Methods of Analysis: SPME–HPLC–MS

It was perhaps logical to assume that after the initial development of SPME for GC that attention would also focus on the use of SPME with HPLC or LC–MS. However, unlike in GC where the injector provides the means for thermal desorption of compounds from the fibre, no such situation exists for LC. For LC therefore, compounds are desorbed from the fibre using the mobile phase, i.e. solvent desorption. This required the development of a separate interface, as described above. Initial work, reported in 1995 [42], focused on the interfacing of SPME with HPLC using the separation and identification of PAHs. The interface device was designed using a standard HPLC instrument incorporating a desorption chamber located in the position usually occupied by the injection loop of a 6-port injection valve. The desorption chamber was made of a 0.75 mm i.d. stainless-steel 'tee' with two of the three ports connected to the injection loop ports of the injection valve. In this work, a 7 µm polydimethylsiloxane fibre was exposed to a stirred water sample spiked with PAHs for 30 min. A comparison between a direct 1 µl loop injection and a fibre injection using 7 µm polydimethylsiloxane extraction for 30 min from a 100 ppb solution of each

PAH was made. It was observed that some 'fibre-selectivity' had occurred for a number of the peaks separated, i.e. acenaphthylene, fluorene, phenanthrene and anthracene. Since its first introduction in 1995 to date, the practical application of SPME–HPLC has lagged behind that of SPME–GC [43]. A number of reasons exist why the SPME–HPLC method has not been widely implemented. These include the small selection of commercially available SPME sorbents, long equilibration times, more challenging desorption optimization, a lack of automation of the methods, the significantly more tedious nature of HPLC desorptions of fibres relative to GC desorptions and lack of commercially available interfacing options. The author has also noted on the interfacing issue that it requires significant modification of the LC injector, whereas the design of conventional injectors does not lend itself to such modification. To date, several options have been applied for SPME–HPLC interfacing but no single strategy or interface device design has proven optimal [43]. The most common configurations available include: (1) use of a manual injection interface 'tee', (2) 'in-tube' SPME and (3) off-line desorption followed by conventional liquid injection. In addition, several experimental set-ups for direct introduction of an SPME fibre via 'electronanospray' to mass spectrometry have been recently discussed [44].

4.5.1 Analysis of Abietic Acid and Dehydroabietic Acid in Food Samples

An investigation of 'in-tube' SPME coupled to liquid chromatography–mass spectrometry (LC–MS) for the analysis of abietic acid and dehydroabietic acid (Figure 4.9) in food samples has been reported [45]. 'In-tube' SPME was invented as a means to completely automate the SPME process [46]. It is similar to the SPME-fibre approach, but the extraction device has an open tubular fused-silica GC capillary column with a proper coating on the internal surface. In this work, a GC capillary column (60 cm × 0.32 mm i.d.) was used as the 'in-tube' SPME

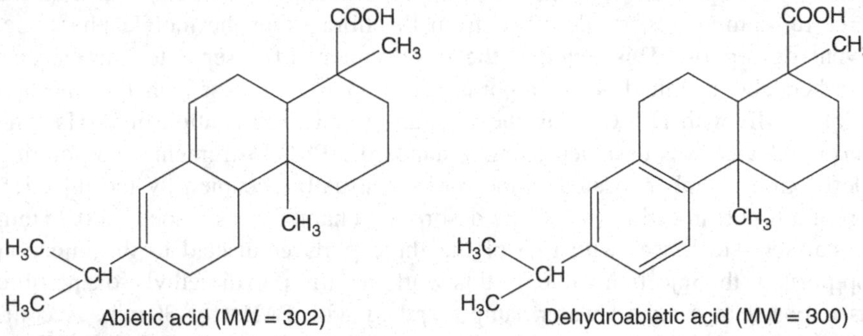

Figure 4.9 Chemical structures of abietic acid and dehydroabietic acid [45].

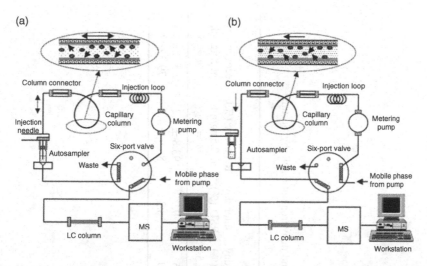

Figure 4.10 Schematic diagrams of the online 'in-tube' SPME–LC–MS system: (a) load position (extraction); (b) injection position (desorption) [45]. Reprinted from *J. Chromatogr., A*, **1146**(1), Mitani *et al.*, 'Analysis of abietic and dehydroabietic acid in food samples by in-tube solid-phase microextraction coupled with liquid chromatography–mass spectrometry', 61–66, Copyirght (2007) with permission from Elsevier.

device, and positioned between the injection loop and injection needle of the autosampler (Figure 4.10). The food samples in liquid form were used directly after filtration with a 0.45 μm syringe microfilter, whereas the semi-solid and solid food samples were dissolved in hot water, followed by centrifugation at 3000 g for 10 min and the supernatant used for the extraction. After 'in-tube' SPME extraction, the compounds were desorbed from the capillary coating and transported to the HPLC column (ODS-3 column and 5 mM ammonium formate/acetonitrile, 10:90 vol/vol) by switching the 6-port valve to the injection position. The compounds were detected by the MS system. The method developed provided good linearity, detection limits, recoveries and reproducibilities (Table 4.7). In addition, greater sensitivity (85- and 75-fold for each compound) than the direct injection method (5 μl injection) was obtained. The method was successfully applied to analyse various liquid and solid food samples contacted with paper and able to detect the compounds at ng ml^{-1} or ng g^{-1} levels without interference peaks.

4.5.2 Analysis of Fungicides in Water Samples

The use of SPME coupled to HPLC with fluorescence detection for extraction and determination of benzimidazole fungicides (benomyl, carbendazim, thiabendazole and fuberidazole) in water has been reported [47]. The optimized conditions were:

Table 4.7 Linearity, detection limits, reproducibilities and recoveries of abietic acid and dehydroabietic acid by 'in-tube' SPME–LC–MS [45]

Compound	Linear range (ng/ml)	Correlation coefficient	Detection limits (pg/ml)		Intra-day RSD (%), $n = 5$	Inter-day RSD (%), $n = 5$	Recovery (%), at 0.5 ng/ml, $n = 3$	Recovery (%), at 5 ng/ml, $n = 3$
			Direct injection (5 µl)	'In-tube' SPME				
Abietic acid	0.5–50	0.9999	248	2.9	4.5	9.9	93.8 ± 6.2	93.1 ± 2.3
Dehydroabietic acid	0.5–50	0.9998	153	2.1	5.9	8.3	86.9 ± 4.5	79.3 ± 2.1

Table 4.8 Analytical performance criteria obtained using SPME combined with HPLC–fluorescence detection [47]

Compound	Linear range (ng/ml)	Correlation coefficient	Detection limits (ng/ml)	RSD (%), $n = 6$
Carbendazim/benomyl	2–300	0.992	1.30	9.0
Thiabendazole	0.5–300	0.999	0.04	6.6
Fuberidazole	0.05–5	0.994	0.03	7.9

SPME fibre, CAR-PDMS 75 μm; extraction time, 40 min; ionic strength, 15% wt/vol NaCl; extraction temperature, 60°C; stirring speed, 600 rpm; desorption time, 10 min. The HPLC separation column used was a 3.9 mm × 150 mm, 8 μm particle diameter, Symmetry C-18. Methanol–water (45:55 vol/vol) at a flow rate of 1.0 ml/min was served as the isocratic mobile phase. In this work, it is noted that the SPME desorption was carried out off-line followed by conventional liquid injection to the HPLC system with a scanning fluorescence detector. The analytical performance of the system is summarized in Table 4.8. The method developed was used for determination of the fungicide compounds in different environmental water samples (sea, sewage and ground waters).

4.6 Automation of SPME

Automation of an analytical method facilitates practical application of the method to routine analysis, especially where sample throughput is high, and it also provides greater reproducibility. The automation of SPME analysis was first published in 1992 [48]. In this work, a Varian model 8100 syringe autosampler was adapted to hold the SPME device. At that time, magnetic stirring was used for agitation and later in 1996 [49] it was replaced by a modified device that allowed vibration of the fibre to agitate the sample. In 1999, CTC Analytics (Zwingen, Switzerland) launched the CombiPAL™ autosampler (Figure 4.11) which has capabilities of full temperature control of individual samples, stirring, fibre conditioning and 'baking out' of the fibre outside the injection port [49]. Automated SPME methods have been applied for the analysis of a variety of compounds [50–55]. Aside from the original fibre-type SPME, the 'in-tube' SPME device has been automated and commercially available since 2000 as a 'solid-phase dynamic extraction' (SPDE) system [56]. An illustration of a sample preparation using SPDE is shown in Figure 4.12. The SPDE method has some limitations in terms of its complexity and requiring a large number of precise plunger strokes; hence it is much better suited to automated methodology and could not be performed as easily in a manual mode [49]. A number of applications of the method have been published [57–63].

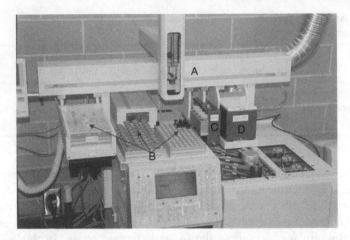

Figure 4.11 Commercial SPME–GC autosampler (CTC Analytics CombiPAL): A, sample preparation/injection arm; B, sample trays; C, needle heater; D, heater/agitator [49]. O'Reilly *et al.*, 'Automation of solid phase microextraction', *J. Sepn Sci.*, 2005, **28**, 2010–2022. Copyright Wiley-VCH Verlag Gmbh & Co. KGaA. Reproduced with permission.

4.6.1 Applications of Automated SPME

4.6.1.1 Analysis of PAHs in Sediments [54, 64]

Recently, a fully automated SPME method has been reported for the analysis of PAHs in sediments at very low levels [54]. This approach involved the use of pressurized hot water extraction (PHWE) followed by SPME and GC–MS analysis. A Dionex ASE-200 extractor was used for the PHWEs. The optimized parameters for PHWEs included an organic modifer (methanol), percentage of organic modifier (10%), temperature (200°C), and static extraction time (10 min). For SPME optimization, the parameters studied had been reported elsewhere [64]: extraction temperature and time (60°C, 60 min), desorption temperature and time (300°C, 10 min), splitless time (4 min), ionic strength (ionic strength correction was not used because the addition of NaCl shortens the lifetime of the fiber) and effect of organic modifier (no organic modifier added). The SPME fibre used was a 65 μm PDMS/DVB. Fully automated SPME was performed by a commercial autosampler CombiPAL connected to the GC–MS system, equipped with an accessory that allowed sample agitating during extraction and fibre cleaning between extractions. The procedure was validated by two standard reference materials (SRM 1944, New York/New Jersey waterway sediment and SRM 1941b, organics in marine sediments). The chromatogram of an extract of SRM 1941b analysed by the PHWE–SPME–GC–MS method is shown in Figure 4.13. The analysis results of the two SRMs (Table 5.9) indicated that the method provided good recovery

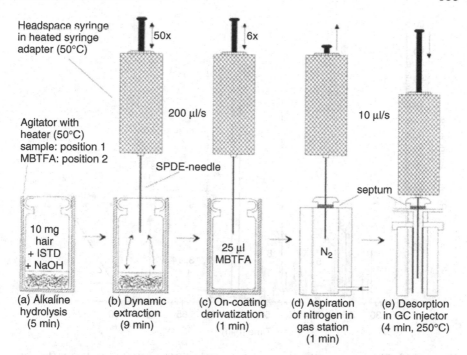

Headspace syringe in heated syringe adapter (50°C)

50x 6x

Agitator with heater (50°C) sample: position 1 MBTFA: position 2

200 µl/s

SPDE-needle

10 µl/s

septum

10 mg hair + ISTD + NaOH

25 µl MBTFA

N₂

(a) Alkaline hydrolysis (5 min)

(b) Dynamic extraction (9 min)

(c) On-coating derivatization (1 min)

(d) Aspiration of nitrogen in gas station (1 min)

(e) Desorption in GC injector (4 min, 250°C)

Figure 4.12 An example of a sample preparation procedure using SPDE with 'in-tube' derivatization [57]. Reprinted from *J. Chromatogr., A*, **958**(1/2), Musshoff *et al.*, 'Automated headspace solid-phase dynamic extraction for the determination of amphetamines and synthetic designer drugs in hair samples', 231–238, Copyright (2002) with permission from Elsevier.

and precision for most of the compounds studied. The calculated limits of detection for the PAHs ranged from 0.4 to 15 µg kg^{-1} and the linearity ranged between 2.5 and 500 µg kg^{-1}. Then, the procedure was applied to the analysis of PAHs at ultratrace levels in sediment samples and proved to be a very promising environmental friendly alternative to the classical methods for the extraction of solid matrices.

4.6.1.2 Determination of Ochratoxin A in Human Urine [65]

Ochratoxin A is produced by some species of *Aspergillus* and is found mainly in tropical regions [65]. It has nephrotoxic, carcinogenic and immunosuppressive properties, and its occurrence in food and feed has been reported worldwide [65]. An automated method using SPME–LC–MS/MS has been applied for the determination of Ochratoxin A in human urine [65]. The approach used an automated multi-fibre system (PAS Technologies, Germany) consisting of a three-arm robotic autosampler and two orbital agitators. Three types of coating were

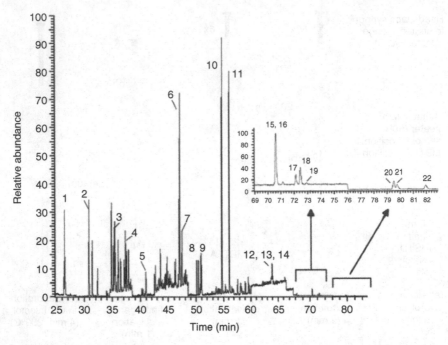

Figure 4.13 Chromatogram of an extract of SRM 1941b analysed by the PHWE–SPME–GC–MS method: 1, naphthalene; 2, methylnaphthalene; 3, acenaphthylene; 4, acenaphthene; 5, fluorine; 6, phenanthrene; 7, anthracene; 8, 1-methylphenanthrene; 9, 2-methylanthracene; 10, fluoranthene; 11, pyrene; 12, benz[*a*]anthracene; 13, triphenylene; 14, chrysene; 15, benzo[*b* + *j*]fluoranthene; 16, benzo[*k*]fluoranthene; 17, benzo[*e*]pyrene; 18, benzo[*a*]pyrene; 19, perylene; 20, dibenz[*a*,*h*]anthracene; 21, benzo[*ghi*]perylene; 22, indeno[1,2,3-*cd*]pyrene [54]. Reprinted from *J. Chromatogr., A*, **1196–1197**(1), Fernández-González *et al.*, 'Pressurized hot water extraction coupled to solid-phase microextraction–gas chromatography–mass spectrometry for the analysis of polycyclic aromatic hydrocarbons in sediments', 65–72, Copyright (2008) with permission from Elsevier.

compared for their extraction efficiency: (1) a C18 coating, (2) a C18/carbon-tape coating and (3) a carbon-tape coating (introduced for the first time in this publication). The carbon-tape coating showed the best extraction efficiency and was chosen for the developed method. The optimized SPME extraction parameters include the following: extraction temperature and time (ambient temperature, 60 min), desorption temperature and time (ambient temperature, 15 min), agitation (850 rpm) and desorption solvent (methanol). It was found that the limits of detection and quantitation were 0.3 and 0.7 ng ml^{-1} in urine, respectively. In addition, the authors claimed that the method for determination of Ochratoxin A meets

Table 4.9 Analysis of the standard reference materials, value ± uncertainty [54]^a. Reprinted from *J. Chromatogr., A*, **1196–1197**(1), Fernández-González *et al.*, 'Pressurized hot water extraction coupled to solid-phase microextraction–gas chromatography–mass spectrometry for the analysis of polycyclic aromatic hydrocarbons in sediments', 65–72, Copyright (2008) with permission from Elsevier

Compounds	SRM 1944			SRM 1941b		
	Certified (mg kg^{-1})	Concentrations (mg kg^{-1})	%R	SRM 1941b (mg kg^{-1})	Concentrations (mg kg^{-1})	%R
Naphthalene	1.65 ± 0.31	1.98 ± 0.51	120	0.848 ± 0.095	0.880 ± 0.110	104
2-Methylnaphthalene	0.95 ± 0.05^b	0.86 ± 0.07	90	0.276 ± 0.053^b	0.247 ± 0.060	90
Acenaphthylene				0.053 ± 0.006	0.052 ± 0.010	97
Acenaphthene	0.57 ± 0.03^b	0.50 ± 0.13	87	0.038 ± 0.005^b	0.039 ± 0.010	102
Fluorene	0.85 ± 0.03^b	0.70 ± 0.04	81	0.085 ± 0.015^b	0.086 ± 0.020	102
Dibenzothiophene	0.62 ± 0.01^b	0.68 ± 0.03	110			
Phenanthrene	5.27 ± 0.22	4.81 ± 1.00	91	0.406 ± 0.044	0.416 ± 0.040	103
Anthracene	1.77 ± 0.33	1.74 ± 0.35	98	0.184 ± 0.018	0.202 ± 0.020	110
1-Methylphenanthrene	1.70 ± 0.10^b	1.97 ± 0.12	116	0.073 ± 0.006^b	0.079 ± 0.010	108
2-Methylanthracene	0.58 ± 0.04^b	0.69 ± 0.06	120	0.036 ± 0.015^b	0.040 ± 0.020	110
Fluoranthene	8.92 ± 0.32	8.17 ± 1.80	92	0.651 ± 0.050	0.609 ± 0.050	94
Pyrene	9.70 ± 0.42	8.04 ± 1.52	83	0.581 ± 0.039	0.554 ± 0.040	95
Benz[a]anthracene	4.72 ± 0.11	4.93 ± 0.16	104	0.335 ± 0.025	0.332 ± 0.030	99
Triphenylene	1.04 ± 0.27	0.98 ± 0.27	95	0.108 ± 0.005	0.112 ± 0.010	104
Chrysene	4.86 ± 0.10	4.39 ± 0.25	90	0.291 ± 0.031	0.283 ± 0.030	97
Benzo[b + j]fluoranthene	5.96 ± 0.40	5.68 ± 0.54	95	0.670 ± 0.021	0.547 ± 0.030	82
Benzo[k]fluoranthene	2.30 ± 0.20	2.28 ± 0.26	99	0.225 ± 0.018	0.174 ± 0.020	77
Benzo[e]pyrene	3.28 ± 0.11	3.37 ± 0.15	103	0.325 ± 0.025	0.314 ± 0.030	97
Benzo[a]pyrene	4.30 ± 0.13	4.20 ± 0.15	98	0.358 ± 0.017	0.246 ± 0.020	69
Perylene	1.17 ± 0.24	1.36 ± 0.24	116	0.397 ± 0.045	0.351 ± 0.050	88
Dibenz[a,h]anthracene	0.42 ± 0.07	0.46 ± 0.08	109	0.053 ± 0.010	0.055 ± 0.010	103
Benzo[ghi]perylene	2.84 ± 0.10	1.56 ± 0.10	55	0.307 ± 0.045	0.100 ± 0.050	32
Indeno[1,2,3-cd]pyrene	2.78 ± 0.10	2.31 ± 0.10	83	0.341 ± 0.057	0.126 ± 0.060	37

Analytical recovery (%, $n = 3$).
^a Expanded uncertainty at the 95% of confidence.
^b Reference values.

the regulatory requirements (as validated according to the 'Food and Drug Administration Guidelines for Bioanalytical Method Validation' in terms of method accuracy, recovery, precision and linearity), and is simpler, less time-consuming and cheaper than other commonly adopted sample clean-up procedures.

SAQ 4.3

It is an important transferable skill to be able to search scientific material of importance to your studies/research. Using your University's Library search engine search the following databases for information relating to the extraction techniques described in this chapter and specifically the use of solid phase microextraction. Remember that often these databases are 'password-protected' and require authorization to access. Possible databases include the following:

- Science Direct;

- Web of Knowledge;

- The Royal Society of Chemistry.

(While the use of 'google' will locate some useful information please use the above databases.)

Summary

The role of solid phase microextraction in recovering organic compounds, either directly from aqueous samples or from the headspace above the samples, is described. The key variables in using solid phase microextraction are highlighted and their applications reviewed. The practical aspects of coupling solid phase microextraction to either gas chromatography or high performance liquid chromatography are described.

References

1. Eisert, R. and Levsen, K., *J. Chromatogr., A.*, **733**, 143–157 (1996).
2. Louch, D., Motlagh, S. and Pawliszyn, J., *Anal.* Chem., **64**, 1187–1199 (1992).
3. Wardencki, W., Curylo, J. and Namiesnik, J., *J. Biochem. Biophys. Meth.*, **70**, 275–288 (2007).
4. Psillakis, E., Ntelekos, A., Mantzavinos, D., Nikolopoulos and Kalogerakis, E., *J. Environ. Monitor.*, **5**, 135–140 (2003).
5. King, A. J., Readman, J. W. and Zhou, J. L., *Anal. Chim. Acta*, **523**, 259–267 (2004).
6. Globig, D. and Weickhardt, C., *Anal. Bioanal. Chem.*, **381**, 656–659 (2005).

7. Mohammadi, A., Yamini, Y. and Alizadeh, N., *J. Chromatogr., A*, **1063**, 1–8 (2005).
8. Ouyang, G., Chen, Y. and Pawliszyn, J., *Anal. Chem.*, **77**, 7319–7325 (2005).
9. Zhang, Z. and Pawliszyn, J., *Anal. Chem.*, **65**, 1843–1852 (1993).
10. Niri, V. H., Bragg, L. and Pawliszyn, H., *J. Chromatogr., A*, **1201**, 222–227 (2008).
11. Lee, M., Chang, C. and Dou, J., *Chemosphere*, **69**, 1381–1387 (2007).
12. van der Wal, L., van Gestel, C. A. M. and Hermens, J. L. M., *Chemosphere*, **54**, 561–568 (2003).
13. Pino, V., Ayala, J. H., Afonso, A. M. and Gonzalez, V., *Anal. Chim. Acta*, **477**, 81–91 (2003).
14. Lambropoulou, D. and Albanis, T., *Anal. Chim. Acta*, **514**, 125–130 (2004).
15. Hawthorne, S. B., Grabanski, C. B., Miller, D. J. and Kreitinger, J. P., *Environ. Sci. Technol.*, **39**, 2795–2803 (2005).
16. Monteil-Rivera, F., Beaulieu, C. and Hawari, J., *J. Chromatogr., A*, **1066**, 177–187 (2005).
17. Bowland, A. and Pawliszyn, J., *J. Chromatogr., A*, **704**, 163–172 (1995).
18. Doong, R. and Liao, P, *J. Chromatogr., A*, **918**, 177–188 (2001).
19. Eriksson, M., Faldt, J., Dalhammar, G. and Borg-Karlson, A.-K., *Chemosphere*, **44**, 1641–1648 (2001).
20. Chia, K., Lee, T. and Huang, S., *Anal. Chim. Acta*, **527**, 157–162 (2004).
21. Navalon, A., Prietol, A., Araujo, L. and Vilchez, J. L., *Anal. Bioanal. Chem.*, **379**, 1100–1105 (2004).
22. Chang, S. and Doong, R., *Chemosphere*, **62**, 1869–1878 (2006).
23. Herbert, P., Morais, S., Paiga, P., Alves, A. and Santos, L., *Anal. Bioanal. Chem.*, **384**, 810–816 (2006).
24. Zuliani, T., Lespes, G., Milacic, R., Scancar, J. and Potin-Gauteir, M., *J. Chromatogr., A*, **1132**, 234–240 (2006).
25. Fernandez-Alvarez, M., Llompart, M., Lamas, J., Lores, M., Garcia-Jares, C., Cela, R. and Dagnac, T., *J. Chromatogr., A*, **1188**, 154–163 (2008).
26. Pawliszyn, J., *Trends Anal. Chem.*, **14**, 113–122 (1995).
27. Ezquerro, O., Ortiz, G., Pons, B. and Tena, M., *J. Chromatogr., A*, **1035**, 17–22 (2004).
28. Perez-Trujillo, J. P., Frias, S., Conde, J. E. and Rodriguez-Delgado, M. A., *J. Chromatogr., A*, **963**, 95–105 (2002).
29. Wu, J., Tragas, C., Lord, H. and Pawliszyn, J., *J. Chromatogr., A*, **976**, 357–367 (2002).
30. Li, H. P., Li, G. C. and Jen, J. F., *J. Chromatogr., A*, **1012**, 129–137 (2003).
31. Goncalves, C. and Alpendurada, M. F., *J. Chromatogr., A*, **1026**, 239–250 (2004).
32. Dong, C., Zeng, Z. and Yang, M., *Water Res.*, **39**, 4204–4210 (2005).
33. Sanchez-Ortega, A., Sampedro, M. C., Unceta, N., Goicolea, M. A. and Barrio, R. J., *J. Chromatogr., A*, **1094**, 70–76 (2005).
34. Campillo, N., Penalver, R. and Hernandez-Cordoba, M., *Talanta*, **71**, 1417–1423 (2007).
35. Junior, J. and Re-Poppi, N., *Talanta.*, **72**, 1833–1841 (2007).
36. Fidalgo-Used, N., Centineo, G., Blanco-Gonzalez, E. and Sanz-Medel, A., *J. Chromatogr., A*, **1017**, 35–44 (2003).
37. Jaber, F., Schummer, C., Al Chami, J., Mirabel, P. and Millet, M., *Anal. Bioanal. Chem.*, **387**, 2527–2535 (2007).
38. Altaki, M. S., Santos, F. J. and Galceran, M. T., *J. Chromatogr., A*, **1146**, 103–109 (2007).
39. 'Determination of Furan in Foods', US Food and Drug Administration (FDA), Washington, DC, USA, 2005 [http://www.cfsan.fda.gov/~dms/furan.html] (accessed, February 2009).
40. Alvarez, I., Bermejo, A. M., Tabernero, M. J., Fernandez, P. and Lopez, P., *J. Chromatogr., B*, **845**, 90–94 (2007).
41. Wejnerowska, G., Karczmarek, A. and Gaca, J., *J. Chromatogr., A*, **1150**, 173–177 (2007).
42. Chen, J. and Pawliszyn, J., *Anal. Chem.*, **67**, 2530–2533 (1995).
43. Lord, H. L., *J. Chromatogr., A*, **1152**, 2–13 (2007).
44. Walles, M., Gu, Y., Dartiguenave, C., Musteata, F. M., Waldron, K., Lubda, D. and Pawliszyn, J., *J. Chromatogr., A*, **1067**, 197–205 (2005).

45. Mitani, K., Fujioka, M., Uchida, A. and Kataoka, H., *J. Chromatogr., A*, **1146**, 61–66 (2007).
46. Eisert, R. and Pawliszyn, J., *Anal. Chem.*, **69**, 3140–3147 (1997).
47. Monzon, A. L., Moreno, D. V., Padron, M. E. T., Ferrera, Z. S. and Rodriguez, J. J. S., *Anal. Bioanal. Chem.*, **387**, 1957–1963 (2007).
48. Arthur, C. L., Killam, L. M., Buchholz, K. D., Pawliszyn, J. and Berg, J. R., *Anal. Chem.*, **64**, 1960–1966 (1992).
49. O'Reilly, J., Wang, Q., Setkova, L., Hutchinson, J. P., Chen, Y., Lord, H. L., Linton, C. M. and Pawliszyn, J., *J. Sepn Sci.*, **28**, 2010–2022 (2005).
50. Frost, R. P., Hussain, M. S. and Raghani, A. R., *J. Sepn Sci.*, **26**, 1097–1103 (2003).
51. Zimmermann, T., Ensinger, W. J. and Schmidt, T. C., *Anal. Chem.*, **76**, 1028–1038 (2004).
52. Mateo-Vivaracho, L., Ferreira, V. and Cacho, J., *J. Chromatogr., A*, **1121**, 1–9 (2006).
53. Luan, T., Fang, S., Zhong, Y., Lin, L., Chan, S. M. N., Lan, C. and Tam, N. F. Y., *J. Chromatogr., A*, **1173**, 37–43 (2007).
54. Fernández-González, V., Concha-Graña, E., Muniategui-Lorenzo, S., López-Mahía, P. and Prada-Rodríguez, D., *J. Chromatogr., A*, **1196–1197**, 65–72 (2008).
55. Vatinno, R., Vuckovic, D., Zambonin, C. G. and Pawliszyn, J., *J. Chromatogr., A*, **1201**, 215–221 (2008).
56. Bicchi, C., Cordero, C., Liberto, E., Rubiolo, P. and Sgorbini, B., *J. Chromatogr., A*, **1024**, 217–226 (2004).
57. Musshoff, F., Lachenmeier, D. W., Kroener, L. and Madea, B., *J. Chromatogr., A*, **958**, 231–238 (2002).
58. Musshoff, F., Lachenmeier, D. W., Kroener, L. and Madea, B., *Forens. Sci. Int.*, **133**, 32–38 (2003).
59. Adbel-Rehim, M., Hassan, Z., Blomberg, L. and Hassan, M., *Therapeut. Drug Monit.*, **25**, 400–406 (2003).
60. Lachenmeier, D. W., Kroener, L., Musshoff, F. and Madea, B., *Rapid Commun. Mass Spectrom.*, **17**, 472–478 (2003).
61. Adbel-Rehim, M., *J. Chromatogr., B*, **801**, 317–321 (2004).
62. Mitani, K., Fujioka, M. and Kataoka, H., *J. Chromatogr., A*, **1081**, 218–224 (2005).
63. Prieto-Blanco, M. C., Cháfer-Pericás, C., López-Mahía, P. and Campíns-Falcó, P., *J. Chromatogr., A*, **1188**, 118–123 (2008).
64. Fernández-González, V., Concha-Graña, E., Muniategui-Lorenzo, S., López-Mahía, P. and Prada-Rodríguez, D., *J. Chromatogr., A*, **1176**, 48–56 (2007).
65. Valenta, H., *J. Chromatogr., A*, **815**, 75–92 (1998).

Chapter 5

New Developments in Microextraction

Learning Objectives

- To appreciate the range of other alternative extraction approaches for recovering organic compounds from aqueous samples.
- To understand the practical aspects of stir-bar sorptive extraction and its applications.
- To understand the practical aspects of single-drop microextraction and its applications.
- To appreciate the diverse range of approaches available for passive sampling of organic compounds in aqueous samples.
- To understand the practical aspects of semipermeable membrane devices for extraction and its applications.
- To be aware of other devices for passive sampling of organic compounds from aqueous samples, namely the polar organic chemical integrative sampler, 'Chemcatcher', ceramic dosimeter and membrane enclosed-sorptive coating device.
- To understand the practical aspects of microextraction in a packed syringe device for extraction and its applications.

5.1 Introduction

A range of different sampling devices have been developed for microextraction of organic compounds from aqueous samples. These are now considered in terms of their method of operation and application.

Extraction Techniques in Analytical Sciences   John R. Dean

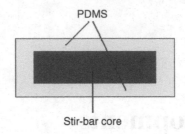

Figure 5.1 Stir-bar sorptive extraction (SBSE).

5.2 Stir-Bar Sorptive Extraction (SBSE)

In the case of stir-bar sorptive extraction (SBSE), organic compounds are pre-concentrated using a magnetic stir bar coated with a sorbent, e.g. polydimethyl-siloxane (PDMS), which is placed in the aqueous sample (Figure 5.1). The stir bar is usually retained in the sample solution (and stirred) for time periods between 30 and 240 min. After the extraction has taken place the stir bar is removed from the solution and gently wiped with a lint-free tissue to remove any retained water droplets. The organic compounds retained on the stir bar (10 mm length × 0.5 mm PDMS coating thickness) then need to be desorbed. This can be carried out by either placing the 'loaded' stir bar in either a small volume of organic solvent and then conventionally injecting the organic compound-containing solvent into either a gas chromatograph or high performance liquid chromatograph or by a thermal desorption unit connected to a gas chromatograph (see Chapter 11, Section 11.2.3). A recent review of SBSE has been published, focusing on its application in environmental and biomedical analysis [1].

5.3 Liquid-Phase Microextraction

5.3.1 Single-Drop Microextraction (SDME)

In single-drop extraction (also known as liquid-phase microextraction, solvent microextraction or liquid–liquid microextraction) a syringe (the same as used for injection of samples in GC – see Chapter 1, Section 5.1) is used to acquire 1 µl of organic solvent (typically toluene due to its low water solubility). This organic solvent is then allowed to exit the syringe but remain as a drop on the end of the needle. The needle is then immersed in the aqueous sample (Figure 5.2). In the case of an aqueous sample, agitation can be achieved by the use of a magnetic stir bar. After a defined period of time (e.g. 30 min) the drop is drawn back into the syringe and then injected into the injection port of a gas chromatograph.

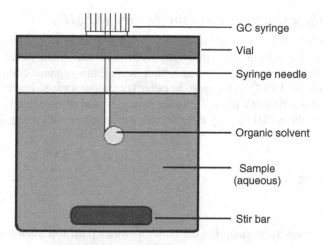

Figure 5.2 Single-drop microextraction.

SAQ 5.1

How might this approach be used with a larger drop of organic solvent?

SAQ 5.2

How might SDME be used for headspace sampling?

The main advantages of this approach are the lack of additional apparatus required (e.g. a gas chromatograph) to achieve rapid extraction and pre-concentration of organic compounds from aqueous samples. The major drawbacks are the selection of an appropriate organic solvent that will form and retain a distinct droplet for extraction, as well as significant manual dexterity on behalf of the analytical scientist. A review of the application of liquid-phase microextraction techniques in pesticide residue analysis has been recently published [2].

5.4 Membrane Microextraction

The use of membrane devices for passive sampling of organic compounds in aqueous samples has developed considerably over recent years. A range of devices has been developed and these are now considered in the following.

5.4.1 Semipermeable Membrane Device (SPMD)

A typical SPMD consists of a low-density polyethylene (LDPE) tubing or membrane. Inside the tubing (or sandwiched between the membrane) is a high-molecular-weight lipid (e.g. triolein) which will retain organic compounds that transfer across the LDPE membrane. In order for this process to occur the organic compounds must be both highly soluble in water and non-ionized. The use of triolein makes the SPMD highly effective for compounds with a log $K_{ow} > 3$ [3].

DQ 5.1

What is log K_{ow}?

Answer

This is a numerical value for the octanol–water partition coefficient that is mathematically logged such that the scale of the number remains small.

5.4.2 Polar Organic Chemical Integrative Sampler (POCIS)

The POCIS consists of a sorbent (receiving phase for organic compounds) positioned between two microporous polyethersulfone diffusion-limiting membranes (Figure 5.3). The choice of sorbent influences the selectivity of the device for organic compounds. A typical sorbent capable of monitoring pesticides is Isolute ENV+, a polystyrene–divinylbenzene copolymer and Ambersorb 1500 carbon dispersed on S-X3 Biobeads.

5.4.3 'Chemcatcher'

The 'Chemcatcher' consists of a 47 mm C18 'Empore' disc (to retain organic compounds, i.e. the receiving phase) and an LDPE diffusion-limiting membrane (Figure 5.3) which are retained within a PTFE housing.

5.4.4 Ceramic Dosimeter

This uses a ceramic tube as the diffusion-limiting barrier which encloses solid sorbent beads (as the receiving phase) (Figure 5.3).

5.4.5 Membrane Enclosed-Sorptive Coating (MESCO) Device

This device consists of a stir-bar sorptive extraction (SBSE) unit (see Section 5.2) as the receiving phase enclosed in a membrane composed of regenerated cellulose as the diffusion-limiting barrier (Figure 5.3).

Several reviews of the applications and developments in membrane extraction have recently been published [3, 5, 6].

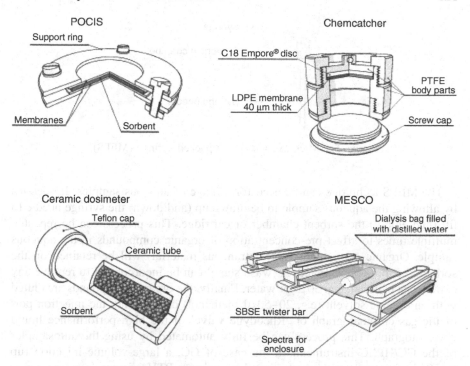

Figure 5.3 Membrane extraction devices for aqueous samples [4]. Reprinted from *Anal. Chim. Acta*, **602**(2), Kot-Wasik *et al.*, 'Advances in passive sampling in environmental studies', 141–163, Copyright (2007) with permission from Elsevier.

5.5 Microextraction in a Packed Syringe (MEPS)

Microextraction in a packed syringe (MEPS) is a new technique for the miniaturization of solid phase extraction. The MEPS device can be directly used instead of a conventional syringe for introduction of samples into a gas chromatograph or high performance liquid chromatograph. In MEPS, a sorbent is located in a chamber (or cartridge) at the top of a syringe needle (Figure 5.4).

DQ 5.2

What types of material could be used as the sorbent?

Answer

Any sorbent that is used for SPE can be used and therefore includes C18, C8, C2, a polystyrene–divinylbenzene copolymer (PS–DVB) or molecularly imprinted polymers (MIPs).

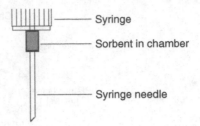

Figure 5.4 Microextraction in a packed syringe (MEPS).

The MEPS technique can be used for a range of aqueous samples. It operates by allowing the aqueous sample to be drawn up (and down) the syringe needle to fill (and empty) the sorbent chamber or cartridge. This process can be repeated multiple times to affect pre-concentration of organic compounds in the aqueous sample. Organic compounds (and extraneous material) will be retained on the sorbent, i.e. pre-concentrated. A 'wash stage' can be incorporated to remove any extraneous material, e.g. 50 μl of water. Finally, the organic compounds are eluted with an organic solvent (e.g. 20–50 μl methanol) directly into the injection port of the gas chromatograph or 'Rheodyne valve' of the high performance liquid chromatograph. This process can be fully automated by using the autosampler of the GC/HPLC instrument. In the case of GC, a large-volume injection (up to 50 μl of extract) can be introduced by using a PTV injector (see Chapter 1, Section 1.5.1). This approach has been applied for the analysis of, for example, PAHs in water [7] and drugs in blood [8].

SAQ 5.3

It is an important transferable skill to be able to search scientific material of importance to your studies/research. Using your University's Library search engine search the following databases for information relating to the extraction techniques described in this chapter and specifically the use of membrane devices used for extraction. Remember that often these databases are 'password-protected' and require authorization to access. Possible databases include the following:

- Science Direct;
- Web of Knowledge;
- The Royal Society of Chemistry.

(While the use of 'google' will locate some useful information please use the above databases.)

Summary

A whole range of alternate approaches for recovering organic compounds from aqueous samples have recently emerged. This chapter describes these new approaches in terms of their instrumentation and application.

References

1. Kawaguchi, M., Ito, R., Saito, K. and Nakazawa, H., *J. Pharm. Biomed. Anal.*, **40**, 500–508 (2006).
2. Lambropoulou, D. A. and Albanis, T. A., *J. Biochem. Biophys. Meth.*, **70**, 195–228 (2007).
3. Vrana, B., Mills, G. A., Allan, I. J., Dominiak, E., Svensson, K., Knutsson, J., Morrison, G. and Greenwood, R., *Trends Anal. Chem.*, **24**, 845–868 (2005).
4. Kot-Wasik, A., Zabiegala, B., Urbanowicz, M., Dominiak, E., Wasik, A. and Namiesnik, J., *Anal. Chim. Acta*, **602**, 141–163 (2007).
5. Barri, T. and Jonsson, J.-A., *J. Chromatogr., A*, **1186**, 16–38 (2008).
6. Esteve-Turrillas, F. A., Pastor, A., Yusa, V. and de la Guardia, M., *Trends Anal. Chem.*, **26**, 703–712 (2007).
7. El-Beqqali, A., Kussak, A. and Abdel-Rehim, M., *J. Chromatogr., A*, **1114**, 234–238 (2006).
8. Abdel-Rehim, M., *LC–GC Eur.*, **22**, 8–19 (2009).

SOLID SAMPLES

Chapter 6

Classical Approaches for Solid–Liquid Extraction

Learning Objectives

- To be aware of approaches for performing solid–liquid extraction of organic compounds from solid samples.
- To understand the principle of operation of Soxhlet extraction and its application.
- To be able to select the most appropriate solvent for Soxhlet extraction.
- To be aware of other approaches for performing solid–liquid extraction and their limitations and benefits: 'Soxtec', sonication and shake-flask.

6.1 Introduction

The extraction of organic compounds, including pesticides, polycyclic aromatic hydrocarbons and phenols from matrices (soils, sewage sludges, vegetables, plants), has historically been carried out by using Soxhlet extraction. Alternate approaches to Soxhlet extraction do exist and include the use of mechanical shaking, often referred to as shake-flask extraction, or ultrasound, in the form of a sonic bath or probe (sonication). While the latter are undoubtedly faster than Soxhlet extraction it is the former which is regarded as the benchmark against which all other approaches are often compared.

The mode of operation of all extraction systems is that organic solvent under the influence of heat (and pressure) will desorb, solvate and diffuse the organic

Extraction Techniques in Analytical Sciences   John R. Dean
© 2009 John Wiley & Sons, Ltd

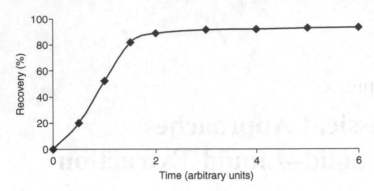

Figure 6.1 Typical extraction profile for the recovery of an organic compound from a solid matrix.

compounds from the sample matrix allowing them to transfer into the bulk (organic) solvent. These processes can be illustrated (Figure 6.1), in the form of the typical two-stage extraction profile.

SAQ 6.1

In Figure 6.1, which extraction process is fast and which is slow?

6.2 Soxhlet Extraction

The apparatus for Soxhlet extraction consists of a solvent reservoir, extractor body, an electric heat source (e.g. an isomantle) and a water-cooled reflux condenser. Two variations of the apparatus are possible: one in which the solvent vapour passes outside (Figure 6.2(a)) or alternatively within the body of the apparatus (Figure 6.2(b)). As the mode of operation of both is the same, only the former will be described in detail.

Soxhlet extraction uses a range of organic solvents to remove organic compounds from predominantly solid matrices.

DQ 6.1

Which solvents might you use for Soxhlet extraction?

Answer

For soil samples, the following solvents are often used: acetone/hexane (1:1, vol/vol); DCM/acetone (1:1, vol/vol); DCM; toluene/methanol (10:1, vol/vol).

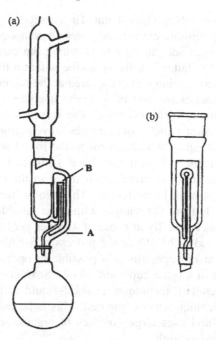

Figure 6.2 Soxhlet extraction processes. (a) Solvent vapour passes external to the sample-containing thimble, which results in cooled organic solvent passing through the sample; this extraction process is relatively slow. (b) Solvent vapour surrounds the sample-containing thimble; the hot organic solvent allows more rapid extraction. From Dean, J. R., *Extraction Methods for Environmental Analysis*, Copyright 1998. © John Wiley & Sons, Limited. Reproduced with permission.

The solid sample is placed in a porous thimble (cellulose) which is located in the inner tube of the extractor body. Often other materials are mixed with the solid samples for specific purposes. For example, to enhance sample–solvent interactions (i.e. maximize the surface area) and reduce sample moisture anhydrous sodium sulfate is added. For samples with high sulfur content, e.g. in the analysis of polycyclic aromatic hydrocarbons in soil sourced from former coal-based power generation plants, the addition of copper powder to the sample in the thimble is required to reduce the possibility of sulfur interference in the subsequent analysis step. The extractor body is then fitted to a round-bottomed flask containing the chosen organic solvent and to a reflux condenser. By heating the solvent with an isomantle (electric heating device) the solvent will gradually become a vapour and pass vertically through the tube marked (A). As the solvent vapour continues to rise it eventually comes into contact with the reflux condenser where the solvent vapour condenses and descends into the extractor body. Within the extractor body is located the sample-containing thimble which now slowly fills with solvent. The passage of warm solvent through the sample-containing thimble extracts any organic compounds contained within it. The extract-containing solvent now rises within the extractor body and also within the 'B' tube. The latter is actually a tube within a tube with the entrance for the rising extract-containing solvent located at the top end. Once the extract-containing solvent reaches the top of the tube it enters the inner tube which is connected to the round-bottomed flask. The solvent entering this inner tube causes a siphoning action which both empties solvent from the extractor body and connecting tubing, all of which returns to the round-bottomed flask. As the extract-containing solvent will normally have a higher boiling point than the original 'pure' solvent it is preferentially retained in the round-bottomed flask, thus allowing 'fresh' solvent to recirculate. This allows 'fresh' solvent to extract the organic compounds from the sample within the thimble. This solvent cycle is repeated many times (typically at a rate of 4 cycles per hour) for several hours (typically between 6 and 24 h). While the process of Soxhlet extraction has been described with one set of apparatus it is possible to operate with as many sets of apparatus as space in a fume cupboard allows. Soxhlet extraction is normally regarded as the 'benchmark technique' in solid–liquid extraction against which all over extraction techniques are compared. This is because, while the process is slow (up to 24 h) and uses large volumes of organic solvent, the extraction recoveries are regarded as high.

DQ 6.2

Which extraction technique is used to recover organic compounds from solid matrices as part of the process of producing certified reference materials (CRMs)? (See Chapter 12, Section 12.2 for details of CRMs.)

Answer

Usually, for the reasons stated, Soxhlet extraction is used to establish the base data on which the certification process is produced.

6.3 Automated Soxhlet Extraction or 'Soxtec'

In 'Soxtec' extraction (Figure 6.3) a three-stage process is used to obtain more rapid extractions than in Soxhlet extraction. In the first stage, a sample-containing thimble is immersed in boiling solvent for approximately 60 min. Then, the sample-containing thimble is removed from the solvent and the process continued as in the Soxhlet extraction approach (see Section 6.1). This second stage is repeated for up to 60 min. In the final stage, solvent evaporation takes place within the Soxtec apparatus, reducing the final extract volume to 1–2 ml

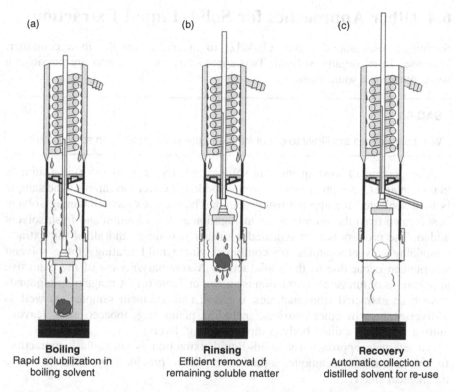

Figure 6.3 'Soxtec apparatus' using a three-step extraction procedure: (a) boiling – extraction of organic compounds occurs by immersing the thimble in boiling solvent; (b) rinsing – thimble containing the sample is raised above the solvent and the process continues as per Soxhlet extraction; (c) recovery – concentration of the sample-containing extract takes place by evaporation, simultaneously collecting the distilled solvent which can be re-used or disposed. Figure drawn and provided by courtesy of Dr Pinpong Kongchana.

in approximately 10–15 min. The advantages of Soxtec over Soxhlet extraction are as follows:

- Rapid extraction (approximately 2 h per sample compared to up to 24 h for Soxhlet extraction).

- Smaller solvent usage (only 20% of the solvent volumes for Soxhlet extraction).

- Sample is concentrated directly within the apparatus.

6.4 Other Approaches for Solid–Liquid Extraction

Sonication uses sound waves (20 kHz) to agitate a sample, in a container, immersed in an organic solvent. Two approaches for sonication are possible: a sonic probe or a sonic bath.

SAQ 6.2

What differences are likely to occur between the sonic probe and sonic bath?

   After placing a known quantity of solid sample (typically 0.5–5 g) in a suitable glass container, enough organic solvent is added to cover the sample. The sample is then sonicated for approximately 3 min. Then, the extract-containing solvent is separated from the sample by centrifugation and/or filtration and fresh solvent added. The process is then repeated a further two times and all of the extract-containing solvent samples are combined. Some mild heating of the solvent/sample can occur due to the sonic action. A summary/review of the extensive applications of ultrasonic extraction is shown in Table 6.1. A range of compounds have been extracted from matrices, e.g. soil and sediment samples, as well as a diverse range of other matrices, including plants (e.g. tobacco, root, leaves), animal feeds and animal body components (e.g. livers).

   An alternate approach for solid–liquid extraction is shake-flask extraction. In this extraction technique, agitation is either provided by hand or via a mechanical shaker.

SAQ 6.3

What possible actions might a mechanical shaker produce?

   A sample (typically 0.5–5 g) is placed into a suitable glass container and enough organic solvent is added to cover the sample. The sample is then agitated by shaking for approximately 3–5 min. Then, the extract-containing solvent is separated from the sample by centrifugation and/or filtration and fresh solvent is added. The process is then repeated a further two times and all of the extract-containing solvent samples combined. Multiple extractions can be easily carried out by using the shake-flask approach with the aid of mechanical laboratory shakers.

   DQ 6.3

   In most cases of solid–liquid extraction, described above, fresh solvent is introduced into the process either deliberately or by the extraction process itself. Why is this so?

(*continued on p. 138*)

Table 6.1 Selected examples of the use of ultrasonic extraction (USE) in analytical sciences[a]

Compounds	Matrix	Typical recoveries	Comments	Reference
USE on compounds from soil matrices				
Veterinary antibiotics: oxytetracycline (OTC), sulfachloropyridazine (SCP) and tylosin (TYL)	Soil (4 different types/studies)	Recoveries were: 68–85% for SCP in all soils; 58–75% for OTC in sandy soils and 27–51% in clay soils; 74–105% for TYL in sandy soil and 47–61% in clay soil	A combination of USE and vortex mixing using a mixture of methanol, EDTA and 'McIlvaine buffer' at pH 7 as the extractant solution. Extracts cleaned-up using SPE with (a) an anion exchange column to remove soil organic matter and (b) a polymeric resin for retention of compounds. Analysis by HPLC–UV or HPLC–FL (for SCP)	1
Fungicides: vinclozolin and dicloran	Soil	Recoveries >91% using procedure (2); limits of detection between 2 and 3 ng/g	Two approaches evaluated: (1) sample extracted with water containing 5% acetone followed by centrifugation or (2) sample extracted with acetone, then diluted with water to produce a 5% vol/vol content. Analysis by SPME–GC–MS	2
OCPs, including α-, β-, γ- and Δ-hexachlorocyclohexane, heptachlor, aldrin, o,p'-DDE, dieldrin, p,p'-DDE, p,p'-DDT, methoxychlor and mirex	Soil	>88% recoveries for three fortification levels between 15 and 200 µg/kg; typical %RSDs were <6%	Extraction optimized with respect to solvent type, amount of solvent, duration of sonication and number of extraction steps. Optimized conditions were: 2×25 ml of petroleum ether: acetone (1:1, vol/vol) for 20 min. Comparison with shake flask and Soxhlet on real soil samples gave comparable extraction efficiencies. Analysis by GC–ECD	3

(continued overleaf)

Table 6.1 (*continued*)

Compounds	Matrix	Typical recoveries	Comments	Reference
Pesticides, including OCPs, OPPs, pyrethroids, triazine and acetanildine	Soil	Extraction efficiency ranged from 69 to 118% (average, 88%). Pesticides detected in the range 0.05–7.0 μg/kg with good precision (7.5–20.5%, average 13.7%RSD)	Method applied to soil from an intensive horticultural area in Portugal. Pesticides detected in soil included: lindane, dieldrin, endosulfan, endosulfan sulfate, 4,4'-DDE, 4,4'-DDD, atrazine, desethylatrazine, alachlor, dimethoate, chlorpyrifos, pendimethalin, procymidone and chlorfenvinphos. Analysis by GC–MS	4
Polybrominated diphenyl ethers (PBDEs)	Soil	Recoveries ranged from 81 to 104% and RSDs from 1 to 9% for fortification levels in the range 0.05, 0.1, 1 and 10 μg/kg	Samples placed in small glass columns and subjected to USE using 5 ml of ethyl acetate and 2 × 15 min extraction time. Analysis by GC–MS	5
USE on compounds from sediment matrices				
Polycyclic aromatic hydrocarbons (PAHs)	Sediment (marine)	Comparable results and better precision obtained compared to reflux method. >90% recovery from CRMs	USE with *n*-hexane–acetone (1:1, vol/vol) on a dried homogenized sample and CRM SRM 1941a. Comparison with reflux method using methanolic potassium hydroxide. Extracts cleaned-up using a miniaturized silica gel column prior to analysis by GC–MS	6

OCPs, including α-, β-, γ- and Δ-hexachlorocyclohexane, heptachlor epoxide, aldrin, endosulfan I, p,p'-DDE, dieldrin, endrin, p,p'-DDD, endosulfan II, p,p'-DDT, endrin aldehyde, endosulfan sulfate, methoxychlor and endrin ketone	Sediment (marine)	Good recoveries obtained for a fortification level of 50 ng/g. Detection limits (based on a signal:noise ratio of 3) ranged from 0.1 to 1 ng/g, dry weight	Extraction optimized with respect to solvent type, amount of solvent and duration of sonication. Optimized conditions were: 2×5 ml of dichloromethane for 20 min. Analysis by GC–ECD	7
Endocrine disruptors: the herbicides diuron and linuron and their degradation products, namely 3,4-dichloroaniline (3,4-DCA), 1-(3-4-dichlorophenyl) urea (DCPU) and 1-(3,4-dichlorophenyl)-3-methylurea (DCPMU)	Sediment (freshwater)	Recoveries ranged from 59.5–85.1%, except 3,4-DCA which was 29.0%	Analysis by HPLC–DAD gave a linear response over the range 5–100 µg/kg with detection limits in the range 0.6–4.6 µg/kg	8

(continued overleaf)

Table 6.1 (*continued*)

Compounds	Matrix	Typical recoveries	Comments	Reference
USE on compounds from miscellaneous matrices				
Polyphenols, including chlorogenic acid, esculetin, rutin, scopoletin and quercitrin	Tobacco (*Nicotina tobaccum L.*)	Recoveries ranged from 96 to 108% with RSDs from 2.0 to 4.6%	Dynamic USE was used as follows: 6 ml of 0.5% wt/vol ascorbic acid in methanol for 10 min. Sample extracts clean-up using C18 cartridges prior to analysis by HPLC	9
Solanesol	Tobacco leaf	Average recoveries were 98.7%. A wide variation in solanesol content was found with respect to geographic origin (0.20–1.50%)	Optimization of USE and saponification procedure. Sample extracts analysed by HPLC–UV produced a linear range of 3.65–4672 ng with a detection limit of 1.83 ng	10
Anthraquinones	Root of *Morinda citrifolia*	Recoveries found to be solvent-dependant (acetone > acetonitrile > methanol > ethanol); highest recoveries obtained using an ethanol–water mixture	Optimization of USE evaluated with respect to temperature (25, 45 and 60 °C), ultrasonic power, solvent types and compositions of ethanol in ethanol–water mixtures. Similar recoveries to Soxhlet extraction and maceration by USE but in a faster time	11
Pesticides (dimethoate and α-cypermethrin)	Olive branches	Recoveries were 99% for α-cypermethrin and 90% for dimethoate	Optimization of USE evaluated with respect to volume of extractant, extraction time, number of extraction steps and sample weight. Optimized conditions were: 3 × 35 ml of hexane for 2 min (in each step) using 1 g of sample. Sample extracts cleaned-up using florisil SPE prior to analysis	12

Chlorinated pesticides	Bird livers	Good recoveries obtained with precision <10%	USE conditions were: 20 ml of *n*-hexane:acetone (4:1, vol/vol) for 30 min using 1 g of sample. Sample extracts cleaned-up using 40% vol/vol sulfuric acid and analysed using HS–SPME–GC analysis. Detection limits ranged from 0.5 to 1.0 ng/g, wet weight. Method applied to livers of various bird species from Greece	13
Quinoxaline-1,4-dioxides	Animal feeds (porcine, chicken and fish)	Recoveries ranged from 92 to 104% on fortified samples, spiked at 5, 50 and 200 mg/kg, except cyadox (>75%). Precision was in the range 2–13%RSD	USE was carried out using methanol/acetonitrile/water (35:35:30, vol/vol/vol). Sample extracts cleaned-up using Alumina N SPE prior to analysis by HPLC–UV. Method applicable to determination of 'multi-residues' of compounds in feed and cereal samples in the range 1–200 mg/kg	14

a Analytical techniques: HPLC–UV, high performance liquid chromatography with ultraviolet detection; HPLC–FL, high performance liquid chromatography with fluorescence detection; HPLC–DAD, high performance liquid chromatography with diode array detection; GC–ECD, gas chromatography with electron capture detection; GC–MS, gas chromatography–mass spectrometry; SPME–GC–MS, solid phase microextraction coupled with gas chromatography–mass spectrometry; HS–SPME–GC, headspace–solid phase microextraction coupled with gas chromatography.

(*continued from p. 132*)

Answer

Extraction is a competitive partitioning process between the organic compound of interest, the sample matrix and organic solvent. Careful choice of organic solvent with respect to the organic compound of interest allows the partitioning process to be competitive. The introduction of fresh organic solvent allows this competitive partitioning to remain, thus allowing maximum transfer of the organic compound into the solvent. Repeating the process multiple times allows maximum recovery of the organic compound. However, the recovery becomes one of 'diminishing return' against the effort required, i.e. if the process was repeated many times it is likely that up to 100% of the organic compound may be recovered in due course but that the cost of time, effort and use of organic solvent make it impractical to perform this series of extractions. A compromise situation is to use a defined set of extractions to achieve an acceptable extraction. In the case of Soxhlet extraction, pure convenience of operation may make an extraction time of 24 h acceptable whereas in sonication/shake-flask extraction three separate extractions is common practice.

SAQ 6.4

It is an important transferable skill to be able to search scientific material of importance to your studies/research. Using your University's Library search engine search the following databases for information relating to the extraction techniques described in this chapter and specifically the use of ultrasonic extraction. Remember that often these databases are 'password-protected' and require authorization to access. Possible databases include the following:

- Science Direct;
- Web of Knowledge;
- The Royal Society of Chemistry.

(While the use of 'google' will locate some useful information please use the above databases.)

Summary

The classical approach for recovering organic compounds from solid samples, namely Soxhlet extraction, is discussed in this chapter. As well as providing the necessary background to the approach the important practical aspects of the technique are described. In addition, alternative approaches for recovering organic compounds from solid matrices are described, i.e. 'Soxtec', sonication and shake-flask.

References

1. Blackwell, P. A., Lutzhoft, H.-C. H., Ma, H.-P., Halling-Sorensen, B., Boxall, A. B. A. and Kay, P., *Talanta*, **64**, 1058–1064 (2004).
2. Lambropoulou, D. A. and Albanis, T. A., *Anal. Chim. Acta*, **514**, 125–130 (2004).
3. Tor, A., Aydin, M. E. and Ozcan, S., *Anal. Chim. Acta*, **559**, 173–180 (2006).
4. Goncalves, C. and Alpendurada, M. F., *Talanta*, **65**, 1179–1189 (2005).
5. Sanchez-Brunete, C., Miguel, E. and Tadeo, J. L., *Talanta*, **70**, 1051–1056 (2006).
6. Banjoo, D. R., and Nelson, P. K., *J. Chromatogr., A*, **1066**, 9–18 (2005).
7. Vagi, M. C., Petsas, A. S., Kostopoulou, M. N., Karamanoli, M. K. and Lekkas, T. D., *Desalination*, **210**, 146–156 (2007).
8. Boti, V. I., Sakkas, V. A. and Albanis, T. A., *J. Chromatogr., A*, **1146**, 139–147 (2007).
9. Gu, X., Cai, J., Zhu, X. and Su, Q., *J. Sepn Sci.*, **28**, 2477–2481 (2005).
10. Chen, J., Liu, X., Xu, X., Lee, F. S.-C. and Wang, X., *J. Pharmaceut. Biomed. Sci.*, **43**, 879–885 (2007).
11. Hemwimol, S., Pavasant, P. and Shotipruk, A., *Ultrason. Sonochem.*, **13**, 543–548 (2006).
12. Pena, A., Ruano, F. and Mingorance, M. D., *Anal. Bioanal. Chem.*, **385**, 918–925 (2006).
13. Lambropoulou, D. A., Konstantinou, I. K. and Albanis, T. A., *Anal. Chim. Acta*, **573–574**, 223–230 (2006).
14. Wu, Y., Wang, Y., Huang, L., Tao, Y., Yuan, Z. and Chen, D., *Anal. Chim. Acta*, **569**, 97–102 (2006).

Chapter 7

Pressurized Fluid Extraction

Learning Objectives

- To be aware of approaches for performing pressurized fluid extraction of organic compounds from solid samples.
- To understand the theoretical basis for pressurized fluid extraction.
- To understand the practical aspects of pressurized fluid extraction.
- To appreciate an approach for method development when using pressurized fluid extraction.
- To appreciate the different modes of operation of pressurized fluid extraction, including *in situ*/selective PFE.
- To be aware of the practical applications of pressurized fluid extraction.

7.1 Introduction

The development of pressurized fluid extraction (PFE) can be traced back to 1995 when the Dionex Corporation launched the Accelerated Solvent Extraction (ASE) system. Since 1995 the use and application of PFE has expanded considerably. The technique is also referred to as pressurized liquid extraction (PLE) or pressurized solvent extraction (PSE). The confusion in terms to describe this extraction technique does create an issue when using Web-based search engines to identify key literature. The term used throughout this chapter is *pressurized fluid extraction*. The use of this term is justified on the grounds that the United States Environmental Protection Agency (USEPA) adopted the name 'pressurized fluid extraction' in their EPA Method 3545 [1]. The basic principal of PFE

Extraction Techniques in Analytical Sciences   John R. Dean
© 2009 John Wiley & Sons, Ltd

is that organic solvents, at high temperature and pressure, are used to extract compounds from sample matrices. The original USEPA method focuses on the extraction of persistent organic pollutants (POPs) from environmental matrices.

SAQ 7.1

What is a persistent organic pollutant (POP)?

The methodology was first proposed as a method (Method 3545) in Update III of the USEPA SW-846 Methods, 1995 [1]. This USEPA method (3545) was developed for application of PFE to the extraction of the following classes of compounds from solid matrices: bases, neutral species, acids (BNAs); organochlorine pesticides (OCPs); OPPs; chlorinated herbicides; PCBs.

DQ 7.1

What does the acronym OPPs stand for?

Answer

Organophosphorus compounds – a range of organic compounds that includes dichlorvos and diazinon.

Table 7.1 identifies key compounds within each of the classes of organic compounds mentioned above.

The term 'solid matrices' is used to refer to samples of sewage sludge, soil, clays and marine/river sediments. The choice of extraction solvent, as recommended in the USEPA Method 3545 [1], corresponds to the class of compound to be extracted, i.e. for extraction of BNAs and OPPs use dichloromethane/acetone (1:1, vol/vol), for OCPs use acetone/hexane (1:1, vol/vol), for PCBs use hexane/acetone (1:1, vol/vol) and for chlorinated herbicides use an acetone/dichloromethane/phosphoric acid solution (250:125:15, vol/vol/vol).

7.2 Theoretical Considerations Relating to the Extraction Process

Pressurized fluid extraction uses organic solvents at elevated pressures and temperatures to enhance the recovery of organic compounds from environmental, food, pharmaceutical and industrial samples. The use of organic solvents at elevated pressures and temperatures is advantageous compared to their use at atmospheric pressure and room (or near room) temperature as it results in enhanced solubility and mass transfer effects, and disruption of surface equilibria [2].

Table 7.1 Specific compounds highlighted in the USEPA Method 3545 [1]

(a) *Base, Neutral, Acids (BNAs)*

Phenol	Bis(2-chloroisopropyl)ether	4-Nitrophenol
2-Chlorophenol	Isophorone	Dibenzofuran
1,4-Dichlorobenzene	2-Nitrophenol	N-Nitrosodiphenylamine
2-Methylphenol	Bis(chlorethoxy)methane	Hexachlorobenzene
o-Toluidine	1,2,4-Trichlorobenzene	Phenanthrene
Hexachloroethane	4-Chloroaniline	Carbazole
2,4-Dimethylphenol	4-Chloro-3-methylphenol	Pyrene
Bis(2-chloroethyl)ether	Hexachlorocyclopentadiene	Benz[a]anthracene
1,3-Dichlorobenzene	2,4,5-Trichlorophenol	Benzo[b]fluoranthene
1,2-Dichlorobenzene	2-Nitroaniline	Benzo[a]pyrene
2,4-Dichlorophenol	2,4-Dinitrotoluene	Dibenz[a,h]anthracene
Naphthalene	4-Nitroaniline	Nitrobenzene
Hexachlorobutadiene	4-Bromophenyl-phenylether	3-Nitroaniline
2-Methylnaphthalene	Pentachlorophenol	Fluorene
2,4,6-Trichlorophenol	Anthracene	Chrysene
2-Chloronaphthalene	Fluoranthene	Benzo[k]fluoranthene
Acenaphthene	3,3'-Dichlorobenzidine	Indeno[1,2,3-cd]pyrene
Benzo[g,h,i]perylene	Acenaphthylene	4-Chlorophenyl-phenylether

(b) *Organochlorine pesticides (OCPs)*

Alpha BHC	Endosulfan II	Dieldrin
Beta BHC	Endrin aldehyde	p,p'-DDD
Delta BHC	Methoxychlor	p,p'-DDT
Heptachlor epoxide	Gamma BHC-lindane	Endosulfan sulfate
Alpha chlordane	Heptachlor	Endrin ketone
p,p'-DDE	Gamma chlordane	Aldrin
Endrin	Endosulfan I	

(c) *Organophosphorus pesticides (OPPs)*

Dichlorvos	Fenthion	Disulfoton
Demeton O&S	Tetrachlorvinphos	Dimethoate
TEPP	Fensulfothion	Chlorpyrifos
Sulfotep	Azinfos methyl	Parathion ethyl
Diazinon	Mevinphos	Tokuthion
Monocrotophos	Ethoprop	Bolstar
Ronnel	Phorate	EPN
Parathion methyl	Naled	Coumaphos

(d) *Chlorinated herbicides*

2,4-D	Dichloroprop	Dicamba
2,4,5-T	2,4-DB	Dinoseb
Dalapon	2,4,5-TP	

(e) *Polychlorinated biphenyls (PCBs)*

PCB 28	PCB 101	PCB 153
PCB 52	PCB 138	PCB 180

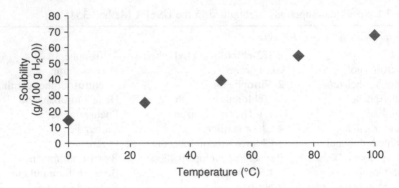

Figure 7.1 Influence of temperature on the solubility of glycine [3].

7.2.1 Solubility and Mass Transfer Effects

As the temperature is increased, the ability of solvents to solubilize compounds also increases. An example of this is given in Figure 7.1 in which the effect of temperature on the solubility of glycine in water is shown.

DQ 7.2

What influence does temperature have on the solubility of glycine?

Answer

It is observed that as the temperature increases so does the solubility of glycine.

In addition, it is also noted that an increase in temperature also leads to faster diffusion rates. Similarly, during the operation of the PFE system (see Section 7.3) fresh solvent is introduced into the system which leads to improved mass transfer of organic compounds from the matrix, i.e. greater extraction rates due to a large concentration gradient between the fresh solvent and the surface of the sample matrix. One of the main benefits of increasing the pressure within the sample cell is that the organic solvents remain liquefied above their (atmospheric pressure) boiling points, thereby promoting solubility effects.

7.2.2 Disruption of Surface Equilibria

The combination of temperature and pressure, in PFE, has concurrent and inter-related benefits which lead to improved recovery of organic compounds from sample matrices. As the temperature within the extraction cell increases it can cause disruption of the strong analyte–matrix interactions caused by hydrogen

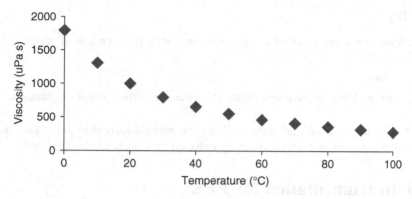

Figure 7.2 Influence of temperature on the viscosity of water [3].

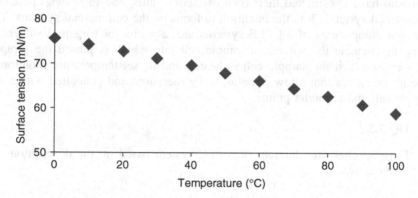

Figure 7.3 Influence of temperature on the surface tension of water [3].

bonding, van der Waals forces and dipole attractions. Also, as the solvent viscosity and surface tension of the organic solvent both decrease as the temperature in the extraction cell is increased (see Figures 7.2 and 7.3, respectively) this allows improved penetration of the solvent within the sample matrix. The resultant affect is that higher extraction efficiencies of the compounds can result.

DQ 7.3

What is the influence of temperature on the viscosity of water?

Answer

It is noted that as the temperature increases the viscosity decreases.

DQ 7.4

What is the influence of temperature on the surface tension of water?

Answer

It is noted that as the temperature increases the surface tension decreases.

The use of a pressurized system allows the organic solvent to penetrate within the sample matrix, thereby promoting enhanced recovery of the analytes.

7.3 Instrumentation for PFE

The instrumentation for PFE can be viewed from two perspectives, those scientists who have constructed their own extraction units, and those who purchase commercial systems. It is the intention to focus on the commercial systems. The common components of all PFE systems are: a source of (organic) solvent, a pump to circulate the solvent, a sample cell into which is placed the sample, an oven in which the sample cell is heated and its set temperature monitored, a series of valves that allow pressure to be measured and generated within the sample cell and an outlet point.

DQ 7.5

Draw a schematic diagram of a PFE system based on the description given above.

Answer

A schematic diagram for a PFE system is shown in Figure 7.4.

The commercial PFE instrumentation is dominated by one supplier (Dionex Corporation) with other systems now beginning to appear on the market. Each system is briefly reviewed in the following.

7.3.1 Dionex System

This PFE system is available in a range of formats, including the ASE® 100, ASE® 200 and ASE® 300 models. The term ASE refers to 'accelerated solvent extractor'. ASE® 100 is a single-cell system whereas the ASE® 200 and 300 systems are automated systems capable of processing 24 or 12 larger samples (>34 ml) sequentially, respectively. The following discussion of the general system will focus on ASE® 200.

A schematic diagram of this PFE system is shown in Figure 7.4. The sample is located in a cell fitted with two finger-tight removable end caps that operate

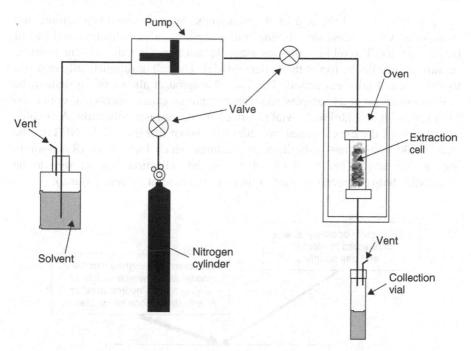

Figure 7.4 Schematic of the layout of a typical pressurized fluid extraction system. From Dean, J. R., *Methods for Environmental Trace Analysis*, AnTS Series. Copyright 2003. © John Wiley & Sons, Limited. Reproduced with permission.

with compression seals and allow high pressure closure. After securing one of the sample cell's end caps onto finger-tightness, a Whatman filter paper (grade D28, 1.98 cm diameter) is introduced inside the cell and gently located by a plunger into the cell's base. Then, the sample and any other associated components (see later) are placed inside the cell. Finally, the other end cap is screwed onto finger-tightness and then the entire sample-containing cell is placed in the carousel. The sample cells range in volume from 0.5, 1, 5, 11, 22 and 33 ml, but all with the same internal diameter of 19 mm. Before performing a pre-specified extraction, an auto-seal actuator places the identified extraction cell into the oven.

In the extraction mode, the sample cell is loaded into the oven, and filled with an appropriate solvent (or solvent mixture) by the solvent supply system. Then, the cell is heated and pressurized for a few minutes (typically 5 min). The ASE® 200 system can operate in the temperature range 40–200°C at pressures of 500–3000 psi (35–200 bar). After completion, the static valves are released and a few ml of fresh solvent are passed through the extraction cell. This process excludes the existing solvent(s) and the majority of the extracted compounds. Then, N_2 gas is purged through the stainless-steel transfer lines and sample cell

(45 s at 150 psi). All the extracted compounds, from an individual sample, are transported via stainless-steel tubing into a septum-sealed collection vial (40 or 60 ml capacity).The tubing contains a needle that punctures the solvent-resistant septum located on the top of the collection vial. The cell is automatically returned to the carousel after extraction. The use of a carousel allows the system to be able to extract up to 24 samples sequentially into an excess number of collection vials (26) with an additional 4 vial positions for rinse/waste collection. A detailed description of the experimental procedure is shown in Figure 7.5. (NOTE: The ASE® 200 system has in-built safety features which include an IR sensor to monitor the arrival and level of solvent in the collection vial, as well as an automatic shut-off procedure that initiates in the case of system failure.)

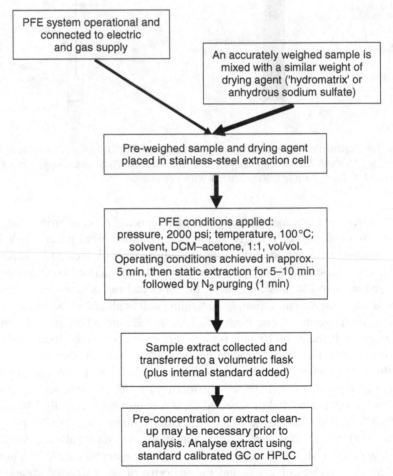

Figure 7.5 Typical analytical procedure used for pressurized fluid extraction.

7.3.2 Applied Separations, Inc.

The Applied Separations (AS) system is commercially available as a pressurized solvent extractor or '*fast* PSE'. It is a fully automated simultaneous extractor which is capable of processing six samples simultaneously. This system is capable of heating samples in the temperature range 50–200°C at pressures of up to 150 bar. Sample cells are available in a range of sizes (11, 22 and 33 ml). (NOTE: The PSE system has in-built safety features to identify leaky fittings, unvented pressure and the absence of an extraction vessel or collection vial. In addition, the AS system is also available as a single-extraction cell system, the '*one* PSE'.)

7.3.3 Fluid Management Systems, Inc.

The Fluid Management Systems (FMS) instrument is commercially available as the pressurized liquid extractor or PLE™. It is a fully automated simultaneous extractor capable of processing between 1 to 6 samples at the same time. An additional benefit of this system is the ability to include an *in situ* sample clean-up module. The FMS system is capable of heating samples in the temperature range 70–200°C at pressures of up to 3000 psi.

7.4 Method Development for PFE

A general approach for preparing and extracting organic compounds from sample matrices is suggested in the following.

General

- In order to assess the integrity of the combined extraction and analysis process it is necessary to establish a benchmark. One approach is to incorporate relevant certified reference materials (CRMs) within the process. The use of CRMs within the overall extraction/analysis protocol allows for an assessment of the accuracy and precision of the procedure; the accuracy being determined by the closeness of the obtained results, and taking into account appropriate errors, against the certified values for the specific and named compounds, whereas the repeated extraction/analysis of the CRM will allow long (and short) term precision, i.e. variability, to be assessed over weeks and months.

Pre-extraction

- Identify and assess the organic compounds to be recovered – this is important in selecting appropriate extraction solvents. Are the compounds soluble in the proposed extraction solvent(s)?

- What is the sample matrix? Wet or moisture-laden samples may need to be either pre-dried or that a moisture-removing adsorbent is added into the extraction cell along with the sample.

- Sample particle size. The smaller the sample particle size, the greater the interaction with the extraction solvent. On that basis it may be appropriate to grind and sieve the sample if it is a convenient form. Alternatively, the sample may need to be freeze-dried prior to grinding and sieving. The reduced particle size combined with enhanced extraction temperatures and pressure will lead to optimum recoveries.

Packing the extraction cell

- How much sample do I have? What size of extraction cell should I use? On the basis of your answers you can proceed.

- Locate a Whatman filter paper in the bottom of the extraction cell using the plunger.

- How should the extraction cell be packed with the sample? Examples of cell packing arrangements are shown in Figure 7.6.

  – To maximize sample surface area it is appropriate to mix the sample with a dispersing agent, e.g. 'Hydromatrix' or diatomaceous earth; suggested ratio of 1 part sample to 1 part 'Hydromatrix'.

  – If the sample is wet or moisture laden (examples might include food matrices) it is appropriate to mix the sample with anhydrous sodium sulfate.

  – If the sample contains significant levels of sulfur (often found at high levels in soils from former gas/coal works) it is necessary to add copper or tetrabutylammonium sulfite powder. The addition of copper or tetrabutylammonium sulfite powder 'complex out' the sulfur preventing it from blocking the stainless-steel tubing within the PFE system.

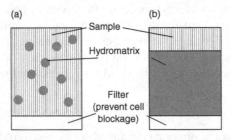

Figure 7.6 Two options for the packing of a PFE cell.

– If the sample is likely to lead to significant co-extractives that could interfere with the post-extraction analysis, e.g. chromatography, it may be opportune to consider an *in situ* sample clean-up using alumina, 'Florisil' or silica gel.

• Finally, ensure that the extraction cell is comfortably full (i.e. remove the dead-volume of the cell). If necessary, add 'Hydromatrix' or similar to remove the void volume.

Extraction conditions

• What extraction conditions within the PFE system is it appropriate to alter? The main operating variables are extraction time (static and dynamic), temperature, pressure and organic solvent. Evidence exists [2] that the majority of compounds are recovered after a 5 + 5 min extraction time. Temperature increases are noted from 50°C up to 100/150°C with little benefit thereafter. Also, you need to consider the potential for compound degradation at elevated temperatures. Similarly pressures of approximately 2000 psi are considered appropriate for recovering most compounds from matrices. In most cases the choice of organic solvents can be considered with respect to the compounds to be extracted. In general, the use of polar solvents will be more effective than non-polar solvents. The recommended solvents, from the USEPA Method 3545 [1], are specifically related to the class of compound to be extracted from sewage sludge, soil, clay and marine/river sediments. For extraction of base, neutral and acid compounds (BNAs) and organophosphorus pesticides (OPPs) a 1:1 vol/vol mixture of dichloromethane/acetone is proposed. While for organochlorine pesticides (OCPs) a 1:1 vol/vol combination of acetone/hexane is proposed; for polychlorinated biphenyls (PCBs) use hexane/acetone (1:1, vol/vol) and for chlorinated herbicides use an acetone/dichloromethane/phosphoric acid solution (250:125:15, vol/vol/vol). It is essential to always use high-purity solvents to minimize chromatographic artefacts.

Maintenance of PFE systems

• Ensure regular maintenance occurs of extraction cells and associated internal fittings and replace, as necessary.

• It is necessary to check the alignment of the collection vial carousel regularly.

• It may be necessary to replace the stainless-steel tubing connection between the extraction cell and collection vial on a periodic basis. The narrow internal diameter of this tubing can become blocked if the sample contains a high sulfur content. As noted above it is possible to alleviate this by the addition of copper powder to the sample pre-extraction.

7.5 Applications of PFE

7.5.1 Parameter Optimization

Any attempt to optimize PFE operating parameters can only be of use if it results in data that have the highest recoveries in the shortest time.

SAQ 7.2

Does it make any sense to attempt any operating parameter optimization when standard conditions are available from the manufacturers and the USEPA?

The main PFE operating parameters considered are as follows:

- Solvent selection or solvent mixtures.

- Optimize static/flush cycles; PFE can perform up to three static–flush cycles in any single extraction.

- Temperature within operational (safe working) limits of 40 and 200°C.

- Pressure within operational (safe working) limits of 1000 and 2400 psi.

- Extraction time within operational (safe working) limits of 2 and 16 min.

The approach to the optimization process also requires some consideration. It is widely regarded that optimization of individual parameters on a 'one-at-a-time' basis is not the most appropriate approach and that a multivariate approach is preferred. However, a significant number of optimization studies undertake the 'one-at-a-time' approach.

Examples of PFE parameter optimization are described in the following.

7.5.1.1 Optimization of PFE: p,p'-DDT and p,p'-DDE from Aged Soils [4]

The influence of solvent and number of extraction cycles on the recovery of DDT and DDE (Figure 7.7) from Ethiopian soils contaminated more than 10 years previously has been investigated. The influence of PFE static extraction time was investigated ($\times$ 10 to $\times$ 40 min) on two different soil samples (labelled A34 and B10) using n-heptane/acetone (1:1, vol/vol) at 100°C (Figure 7.8). It can be seen (Figure 7.8) that approximately 87% DDT and 97% DDE recoveries were obtained in the first 10 min cycle. Additional extraction time up to a 3 $\times$ 10 min cycle allows a cumulative recovery of 97% for DDT and 99% DDE (Note: All recovery data was assessed in terms of recoveries from a 4 $\times$ 10 min cycle). The authors also investigated the influence of a single solvent (n-heptane) and a solvent mixture (n-heptane/acetone, 1:1, vol/vol) on the exhaustive extraction of DDT and DDE from the same soils (Figure 7.9). It is noted that the highest recoveries were obtained using the solvent mixture.

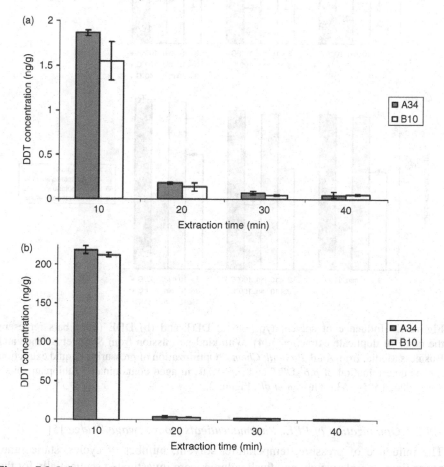

Figure 7.7 Molecular structures of dichlorodiphenyltrichloroethane (DDT) and dichlorodiphenyldichloroethylene (DDE).

Figure 7.8 Influence of the number of extraction cycles on (a) DDT and (b) DDE recoveries (error bars represent the range of duplicate extractions) [4]. With kind permission from Springer Science and Business Media, from *Anal. Bioanal. Chem.*, 'Optimization of pressurized liquid extraction for the determination of p,p'-DDT and p,p'-DDE in aged contaminated Ethiopian soils', **386**, 2006, 1525–1533, Hussen *et al.*, Figure 1.

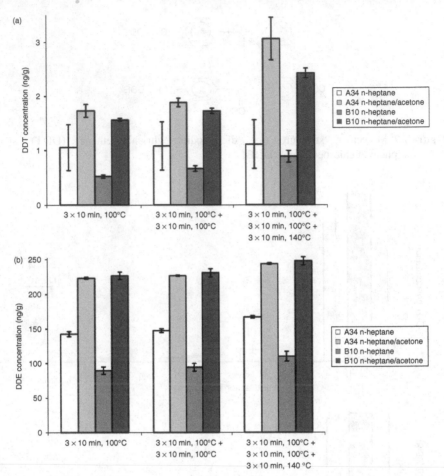

Figure 7.9 Influence of solvent type on (a) DDT and (b) DDE (error bars represent the range of duplicate extractions) [4]. With kind permission from Springer Science and Business Media, from *Anal. Bioanal. Chem.*, 'Optimization of pressurized liquid extraction for the determination of *p,p'*-DDT and *p,p'*-DDE in aged contaminated Ethiopian soils', **386**, 2006, 1525–1533, Hussen *et al.*, Figure 2.

7.5.1.2 Optimization of PFE: Pharmaceuticals from Sewage Sludge [5]

The influence of pressure, temperature, solvent, number of cycles, static time, purge time, sample weight and flush volume were investigated sequentially for the recovery of pharmaceuticals (acetaminophen, caffeine, metoprolol, propanolol, carbamazepine, salicylic acid, bezafibrate, naproxen, clofibric acid, diclofenac and ibuprofen) from spiked sewage sludge. The choice of solvent was investigated first. The solvents investigated were water with 50 mM H_3PO_4/acetonitrile

(9:1, vol/vol), water with 50 mM H_3PO_4/acetonitrile (1:1, vol/vol), water with 50 mM H_3PO_4/acetonitrile (1:9, vol/vol), water with 50 mM H_3PO_4/methanol (1:1, vol/vol), water/methanol (1:1, vol/vol) and water (pH 10)/methanol (1:1, vol/vol). The solvent mixture, water with 50 mM H_3PO_4/methanol (1:1, vol/vol) gave the highest recoveries and was used for further experiments. The next parameter investigated was the number of extraction cycles at a pressure of 1500 psi, a temperature of 100°C, a 15 min static time, a 300 s purge time and 150% flush volume. It was found that the majority of compounds were extracted in the first extraction cycle with some residual extracts in the second cycle and minimal/neglible extracts in the third cycles; two cycles were determined to be the most effective. The flush volume was also investigated; it was found that 150% was the ideal and so its value was continued. A similar process was applied to the purge time, pressure, temperature and static time; it was found that optimum recoveries were obtained. In spite of all of the parameters investigated, the recoveries of salicylic acid were always poor/low. For that reason, salicylic acid was excluded from the study. The methodology was applied to sewage samples from two different sewage treatment farms over a period of 15 months and the data reported.

7.5.1.3 *Optimization of PFE: Sulfonamide Antibiotics from Aged Agricultural Soils* [6]

Sample cells (11 ml) were prepared with 4 g soil and diatomaceous earth and then subjected to the following conditions: extraction solvent at different pH values (2.2, 4.1 and 8.8), temperature (60 to 200°C), extraction time (5 to 99 min) and pressure (100 to 200 bar) to assess recovery of sulfonamide antibiotics from a reference soil (with confirmation from a 'control soil'). In addition, 1 to 3 sequential extractions and a flush volume from 10 to 150% were also tested. The five sulfonamide antibiotics evaluated were sulfadiazine, sulfadimethoxine, sulfamethazine, sulfamethoxazole and sulfathiazole. The major influencing operating variable was assessed to be extraction temperature. All sulfonamides, with the exception of sulfamethoxazole, gave large increases (up to a factor of 6) in recovery when the extraction temperature was increased. A significant issue with sulfonamides is their thermal stability at high temperatures. This was eliminated as an issue by performing spiked experiments on diatomaceous earth at the highest temperatures. It was also noted that the higher-temperature extractions also produced a higher matrix load (visually observed by the darker coloured extracts) which can affect detection by LC–MS/MS. The influence of the matrix on ion suppression was compensated by the use of internal standards. Extraction solvent was assessed and found to be most effective with a mixture of water and acetonitrile (85:15, vol/vol). The influence of pH was also assessed due to the amphoteric nature of the sulfonamides and their expected different interactions with soil. Therefore, the solvent was buffered at pH 2.2 (with formic acid), pH 4.1 (with acetate buffer) and pH 8.8 (with 'Tris buffer'). The highest recovery was obtained with a pH of 8.8. Furthermore it was determined that the other range of operating parameters, i.e. pressure (100 bar), extraction time (9 min pre-heating followed by 5 min static), flush volume (100%) and one extraction cycle did not influence extraction efficiency from the reference soil and so were subsequently used. The developed method was applied to field experiments investigating the fate of sulfonamides after two controlled manure applications.

7.5.2 *In situ Clean-Up or Selective PFE*

One of the strengths of the PFE approach is that within a short time it can effectively recover analytes from matrices. Frequently however, this process is neither selective nor 'gentle'. As a consequence, extraneous material is recovered from the sample matrix which will often interfere with the subsequent analysis step, e.g. chromatography. In order to circumvent this problem, two scenarios are possible. In the 'traditional' approach sample extracts are cleaned-up off-line using, for example, column chromatography or solid phase extraction cartridges containing a particular adsorbent, i.e. alumina, 'florisil' or silica gel. An alternative strategy is to include the adsorbent within the extraction cell along with the sample and perform *in situ* clean-up PFE.

When designing an *in situ* selective PFE approach it is important to think about the following:

- What are your aims when using this approach?
- What do you hope to remove?
- How is it done currently off-line?

Current approaches to perform sample extract clean-up to remove 'chromatographic-interfering components' use one of the following:

- Adsorption: alumina, 'florisil', silica gel.
- Gel permeation chromatography: size separation (removal of high-molecular-weight material).

'**Florisil**' is magnesium silicate with basic properties and allows selective elution of compounds based on elution strength. In contrast, **Alumina** is a highly porous and granular form of aluminium oxide which is available in 3 pH ranges (basic, neutral and acidic). Finally, **Silica gel**, which allows selective elution of compounds based on elution strength. In contrast, **gel permeation chromatography** (GPC) uses a size-exclusion process based on organic solvents and hydrophobic gels to separate macromolecules from the desired compounds.

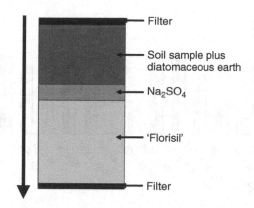

Figure 7.10 An example of how an extraction cell is packed for selective PFE [7].

An example of this approach is the recovery of organochlorine pesticides (OCPs) from a CRM (811-050) and other soils samples using either PFE with off-line clean-up and *in situ* PFE (referred to by the authors of the paper as selective pressurized fluid extraction or SPLE) [7]. In the *in situ* PFE approach the extraction cell (34 ml volume) is packed in the following order (exit point of the cell first): filter, activated 'florisil' (10 g), sodium sulfate (2 g), soil sample mixed with diatomaceous earth (either 4 g of soil or 0.3 g of CRM were mixed with 1 g of diatomaceous earth) and filter (see Figure 7.10). Samples were then extracted as follows: 3 × 10 min at 100°C and 10.4 MPa using 1:1 vol/vol acetone/n-heptane. Extracts were next rotary evaporated to about 1 ml and then quantitatively transferred to GC vials with n-heptane (final volume 1.5 ml). In the case of off-line PFE, samples were extracted under the same experimental conditions, except that the extraction cell did not contain 'florisil' or sodium sulfate. Off-line clean-up was performed as follows: evaporated samples were passed through a column containing activated 'florisil' (4 g) and sodium sulfate (2 g) and then the analytes were eluted with 50 ml of 1:1 vol/vol ethyl acetate/n-heptane. The eluate was rotary evaporated to 0.5 ml and then quantitatively transferred to GC vials with n-heptane (final volume 1.5 ml). The results for the CRM are shown in Figure 7.11. In general terms, the recovery by *in situ* PFE produces slightly lower recoveries (10–20%) than those obtained by off-line PFE. It was postulated that the lower quantity of solvent used in *in situ* PFE (only 17 ml) may have led to less than total recovery from the 'florisil' adsorbent. It is also noted that the average errors were of the order of 10–15% for each approach (typical standard deviations ranged from 1 to 32% for *in situ* PFE, whereas for off-line PFE only they ranged from 1 to 40%). However, all data were within the prediction intervals, of the CRM, provided by the supplier.

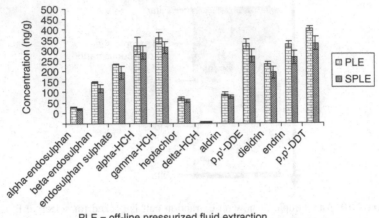

PLE = off-line pressurized fluid extraction
SPLE = *in situ* selective pressurized fluid extraction

Figure 7.11 *In situ* PFE of organochlorine pesticides from a certified reference material (CRM 811-050) ($n = 3$) [7]. Reprinted from *J. Chromatogr., A*, **1152**(1/2), Hussen *et al.*, 'Selective pressurized liquid extraction for multi-residue analysis of organochlorine pesticides in soil', 247–253, Copyright (2007) with permission from Elsevier.

Advantages of *in situ* PFE include the following:

- Increased level of automation of the sample preparation stage.

- Eliminates the need for off-line clean-up.

- Uses less solvent.

- Considerably faster than off-line clean-up.

- Less sample manipulation.

This approach for *in situ* selective PFE based on an in-line clean-up strategy has been applied to a range of sample types and matrices. Other recent examples include: polychlorinated dibenzo-p-dioxins, dibenzofurans and 'dioxin-like' polychlorinated biphenyls from feed and feed samples [8]; polychlorinated biphenyls from fat-containing samples [9]; polychlorinated biphenyls from fat-containing food and feed samples [10, 11]; polycyclic aromatic hydrocarbons and their oxygenated derivatives in soil [12]; polybrominated diphenyl ether congeners in sediment samples [13].

7.5.3 Shape-Selective, Fractionated PFE

A variation on the selective PFE approach described above is shape-selective, fractionated PFE [14]. This approach has been developed for PCBs, PCDDs and

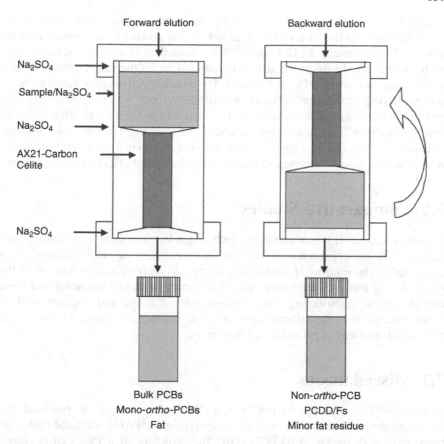

Figure 7.12 'Shape-selective' fractionated pressurized fluid extraction: set-up of the extraction cell [14]. Reprinted from *Trends Anal. Chem.*, **25**(4), Björklund *et al.*, 'New strategies for extraction and clean-up of persistent organic pollutants from food and feed samples using pressurized liquid extraction', 318–325, Copyright (2006) with permission from Elsevier.

PCDFs and involves the insertion of an active carbon column inside the 34 ml PFE cell (Figure 7.12). The PFE system was operated under constant conditions: temperature, 100°C; purge time, 90 s; flush volume, 60%; extraction time, 5 min. Initial work [14] on recovering PCBs, PCDDs and PCDFs from fish oil attempted to extract and fractionate *in situ* within the PFE cell such that bulk PCBs and mono-*ortho* PCBs were collected in a forward elution through the cell whereas non-*ortho* PCBs, PCDDs and PCDFs were eluted in a reverse elution. In the forward elution mode, two fractions were obtained. When using *n*-heptane only (fraction 1) this eluted most of the fat, the bulk PCBs, the mono-*ortho* PCBs and some non-*ortho* PCBs, while the use of a DCM/*n*-heptane (1:1, vol/vol) solvent system (fraction 2) eluted the remaining non-*ortho* PCBs. After stopping the extraction process the PFE cell was turned upside down and re-inserted into the system. The remaining PCDDs and PCDFs were then eluted with toluene only. Some additional clean-up was also required off-line on the toluene fraction prior to determination of PCDDs and PCDFs. By increasing the cell volume to 66 ml and modifying the elution solvents, a revised protocol was proposed [15]. The modified solvent system (and number of cycles) was as follows: fraction 1, 2 × *n*-heptane; fraction 2, 1 × *n*-heptane/acetone (1:2.5 vol/vol); fraction 3, 4× toluene. This revised protocol was applied to an in-house salmon tissue reference sample with excellent results obtained in terms of recovery and effective fractionation.

7.6 Comparative Studies

A comparison of different extraction techniques is often used to assess the performance of one approach over another. Often any new or modified approach is compared to the traditional Soxhlet extraction. As well as a consideration of the recoveries of analytes from matrices, other comparators are necessary and these include capital and running costs, organic solvent usage and operator skill. A fuller description of the different approaches for extraction of organic compounds from solid matrices is provided in Chapter 12.

7.7 Miscellaneous

A study of PFE cell blanks has been undertaken to assess the potential and likely interferences that may arise when analysing for PAHs, aliphatic hydrocarbons and OCPs by GC–FID/ECD [16]. The structure of a PFE cell is shown in Figure 7.13. The evaluation process, using 11 ml cells, was as follows. After reaching the following operating conditions of pressure (2000 psi), temperature (100°C) and solvent (hexane/acetone (1:1, vol/vol)), the cell was maintained under these conditions for 5 min (static extraction). The 'extracts' were then collected, with a rinse stage of fresh solvent, and finally the cell is purged with N_2. Extracts were then concentrated 'to a drop' using a rotary evaporator and to dryness under a stream of N_2. Residues were then reconstituted in 1 ml hexane and analysed by GC–ECD for pesticides and GC–FID for PAHs and aliphatic hydrocarbons. Figure 7.14 shows the GC–FID cell blank scan in comparison to a 0.25 ppm PAH standard scan, indicating the potential interference issues from the blank when analysing for PAHs in PFE sample extracts. A similar problem is highlighted in Figure 7.15 when analysing for pesticides in a soil sample by GC–ECD. A detailed analysis of the cell blank 'extract' was carried out using GC–MSD in full scan mode and the results are shown in Figure 7.16. A range of potential interferents are identified, including silicones and phthalates. A further investigation was performed by microwave extraction of the PEEK ring

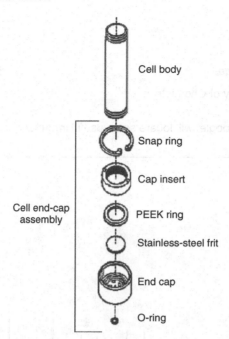

Cell body

Snap ring

Cap insert

Cell end-cap
assembly

PEEK ring

Stainless-steel frit

End cap

O-ring

Figure 7.13 Diagrammatic structure of a PFE cell [16]. With kind permission from Springer Science and Business Media, from *Anal. Bioanal. Chem.*, 'Trouble shooting with cell blanks in PLE extraction' **383**, 2005, 174–181, Fernandez-Gonzalez *et al.*, Figure 1.

using hexane:acetone (1:1, vol/vol). The microwave extract was then analysed using GC–FID and compared with the PFE cell blank extract (Figure 7.17). The similarity of peak retention times between the PFE cell blank extract and the microwave extract of the PEEK seal is noted. It was therefore concluded that the PEEK rings were the most likely source of contaminants. It was proposed that PFE cells must be cleaned prior to analytical use using the same procedure as applied for sample extraction.

SAQ 7.3

It is an important transferable skill to be able to search scientific material of importance to your studies/research. Using your University's Library search engine search the following databases for information relating to the extraction techniques described in this chapter, i.e. pressurized fluid extraction (pressurized liquid extraction or accelerated solvent extraction). Remember that often these databases are 'password-protected' and require authorization to access. Possible databases include the following:

(continued overleaf)

(*continued*)

- Science Direct;
- Web of Knowledge;
- The Royal Society of Chemistry.

(While the use of 'google' will locate some useful information please use the above databases.)

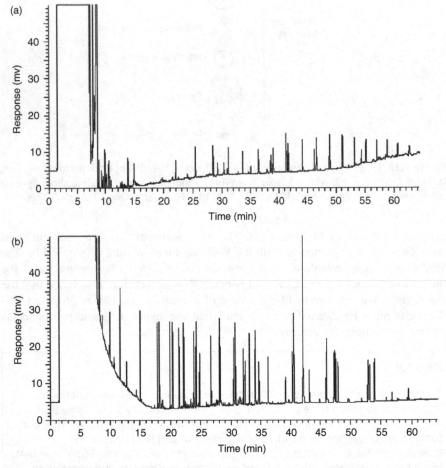

Figure 7.14 GC-FID scans: (a) cell blank; (b) 0.25 ppm PAH standard [16]. With kind permission from Springer Science and Business Media, from *Anal. Bioanal. Chem.*, 'Trouble shooting with cell blanks in PLE extraction', **383**, 2005, 174–181, Fernandez-Gonzalez *et al.*, Figure 2.

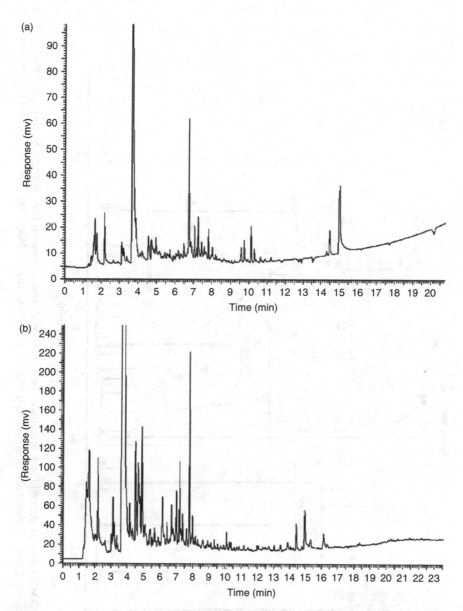

Figure 7.15 GC-ECD scans: (a) cell blank; (b) soil extract [16]. With kind permission from Springer Science and Business Media, from *Anal. Bioanal. Chem.*, 'Trouble shooting with cell blanks in PLE extraction', **383**, 2005, 174–181, Fernandez-Gonzalez *et al.*, Figure 3.

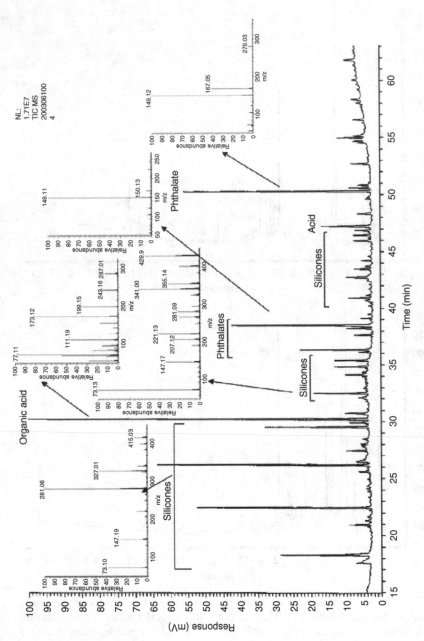

Figure 7.16 GC-MSD scan for the cell blank in full-scan mode [16]. With kind permission from Springer Science and Business Media, from *Anal. Bioanal. Chem.*, 'Trouble shooting with cell blanks in PLE extraction', **383**, 2005, 174–181, Fernandez-Gonzalez *et al.*, Figure 5.

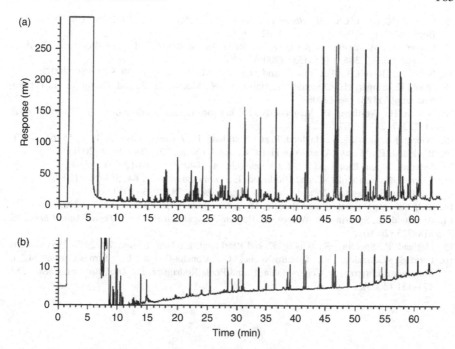

Figure 7.17 GC-FID scans: (a) MAE (1:1, vol/vol hexane/acetone) of the cell peek ring; (b) PFE cell blank [16]. With kind permission from Springer Science and Business Media, from *Anal. Bioanal. Chem.*, 'Trouble shooting with cell blanks in PLE extraction', **383**, 2005, 174–181, Fernandez-Gonzalez *et al.*, Figure 6.

Summary

This chapter describes one of the most important extraction techniques for recovering organic compounds from solid samples, i.e. pressurized fluid extraction. The variables in selecting the most effective approach for pressurized fluid extraction are described. Recent developments in terms of *in situ* clean up/selective extraction, are highlighted and described. The commercial instrumentation for pressurized fluid extraction is also described. A review of the applications of pressurized fluid extraction highlights the diversity of application of this technique.

References

1. USEPA, 'Test Methods for Evaluating Solid Waste', Method 3545, USEPA SW-846, 3rd Edition, Update III, US GPO, Washington, DC, USA (January, 1995).
2. Richter, B. E., Jones, B. A., Ezell, J. L., Avdalovic, N. and Pohl, C., *Anal. Chem.*, **68**, 1033–1039 (1996).

3. Lide, D. R. (Ed.), *CRC Handbook of Chemistry and Physics*, 73rd Edition, CRC Press, Inc., Boca Raton, FL, USA, pp. 6–10 (1992–1993).

4. Hussen, A., Westbom, R., Megersa, N., Retta, N., Mathiasson, L. and Bjorklund, E., *Anal. Bioanal. Chem.*, **386**, 1525–1533 (2006).

5. Nieto, A., Borrull, F., Pocurull, E. and Marce, R. M., *J. Sepn Sci.*, **30**, 979–984 (2007).

6. Stoob, K., Singer, H. P., Stettler, S., Hartmann, N., Mueller, S. R. and Stamm, C. H., *J. Chromatogr., A*, **1128**, 1–9 (2006).

7. Hussen, H., Westbom, R., Megersa, N., Mathiasson, L. and Bjorklund, E., *J. Chromatogr., A*, **1152**, 247–253 (2007).

8. Wiberg, K., Sporring, S., Haglund, P. and Bjorklund, E., *J. Chromatogr., A*, **1138**, 55–64 (2007).

9. Bjorklund, E., Muller, A. and von Holst, C., *Anal. Chem.*, **73**, 4050–4053 (2001).

10. Sporring, S. and Bjorklund, E., *J. Chromatogr., A*, **1040**, 155–161 (2004).

11. Sporring, S., von Holst, C. and Bjorklund, E., *Chromatographia*, **64**, 553–557 (2006).

12. Lundstedt, S., Haglund, P. and Oberg, L., *Anal. Chem.*, **78**, 2993–3000 (2006).

13. de la Cal, A., Eljarrat, E. and Barcelo, D., *J. Chromatogr., A*, **1021**, 165–173 (2003).

14. Bjorklund, E., Sporring, S., Wiberg, K., Haglund, P. and von Holst, C., *Trends Anal. Chem.*, **25**, 318–325 (2006).

15. Haglund, P., Sporring, S., Wiberg, K. and Bjorklund, E., *Anal. Chem.*, **79**, 2945–2951 (2007).

16. Fernandez-Gonzalez, V., Grueiro-Noche, G., Concha-Grana, E., Turnes-Carou, M. I., Muniategui-Lorenzo, S., Lopez-Mahia, P. and Prada-Rodriguez, D., *Anal. Bioanal. Chem.*, **383**, 174–181 (2005).

Chapter 8
Microwave-Assisted Extraction

Learning Objectives

- To be aware of approaches for performing microwave-assisted extraction of organic compounds from solid samples.
- To understand the theoretical basis for microwave-assisted extraction.
- To understand the practical aspects of microwave-assisted extraction.
- To appreciate the potential variables when performing microwave-assisted extraction.
- To be aware of the practical applications of microwave-assisted extraction.

8.1 Introduction

The use of microwaves in analytical sciences is not new. The first reported analytical use for microwave ovens was almost 35 years ago for the digestion of samples for metal analysis [1], with the first use of microwaves for organic compound extraction some ten years later [2]. All microwaves, whether they are found in the home or the laboratory, operate at one frequency, i.e. 2.45 GHz, even though in practice the microwave region exists at frequencies of 100 GHz to 300 MHz (or wavelengths from 0.3 mm to 1 m).

The components of a microwave system are as follows:

- a microwave generator;

- a waveguide for transmission;

Extraction Techniques in Analytical Sciences   John R. Dean
© 2009 John Wiley & Sons, Ltd

- a resonant cavity;

- a power supply.

The microwave generator is called a *magnetron* (Figure 8.1); a phrase first described by A. W. Hull in 1921 [3]. At the microwave frequency (2.45 GHz), electromagnetic energy is conducted from the magnetron to the resonant cavity using a waveguide (or coaxial cable). The sample placed inside the resonant cavity is therefore subjected to microwave energy.

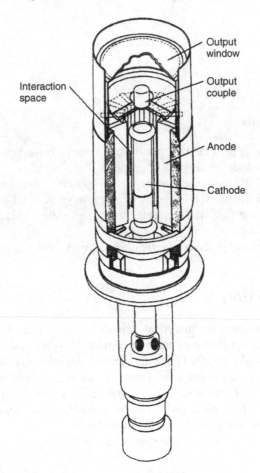

Figure 8.1 Microwave generator: magnetron. Reproduced by permission of Pergamon Press from *Encyclopaedic Dictionary of Physics*, Volume 4, Intermediate State to Neutron Resonance Level, Thewlis, J. (Editor-in-Chief), Pergamon Press, Oxford, UK, p. 486 (1961).

SAQ 8.1

What causes the heating effect of microwaves on samples?

The selection of an organic solvent for microwave-assisted extraction (MAE) is essential; the solvent must be able to absorb microwave radiation and thereby becomes hot. The ability of an organic solvent to be useful for MAE can be assessed in terms of its dielectric constant, ε'; the larger the value of the dielectric constant, the better the organic solvent's ability to become hot. A range of solvents and their respective dielectric constants is shown in Table 8.1.

Table 8.1 Common organic solvents used in MAE [4]

Solvent	Dielectric constant	Boiling point (°C)	Closed-vessel temperature (°C)[a]
Acetone	20.7	56.2	164
Acetonitrile	37.5	81.6	194
Dichloromethane	8.93	39.8	140
Hexane	1.89	68.7	–
Methanol	32.63	64.7	151

[a] At 175 psig.

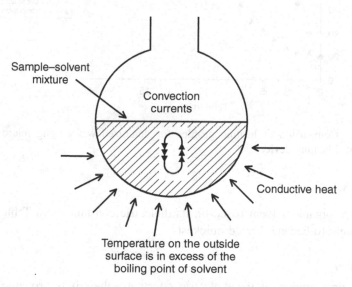

Figure 8.2 Conventional heating of organic solvents. Figure drawn and provided by courtesy of Dr Pinpong Kongchana.

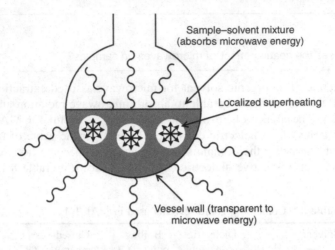

Figure 8.3 Microwave heating of organic solvents. Figure drawn and provided by courtesy of Dr Pinpong Kongchana.

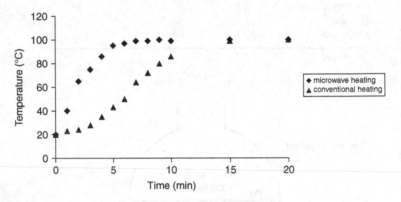

Figure 8.4 Comparison of heating profiles for deionized water using microwave and conventional heating devices.

DQ 8.1

Which organic solvent based on its dielectric constant, from Table 8.1, is likely to become heated quickest?

Answer

By consideration of the dielectric constant values it is probable that acetonitrile with the highest dielectric constant value is likely to be heated quickest (closely followed by methanol).

SAQ 8.2

Why should using a microwave device result in reduced times for extracting organic compounds from sample matrices?

Illustrations of conventional heating and microwave heating of organic solvents are shown in Figures 8.2 and 8.3, respectively, while a comparison of the heating profiles for deionized water using microwave and conventional heating devices is shown in Figure 8.4.

8.2 Instrumentation

Two distinct approaches exist for the use of microwave devices for MAE; one approach uses an open (atmospheric) MAE system (Figure 8.5) whereas the other uses closed (pressurized) MAE (Figure 8.6). In the open (atmospheric) MAE system (Figure 8.5), the sample is located in an 'open vessel' to which an appropriate organic solvent is added. Microwaves are directed via the waveguide onto the sample/solvent system, thus causing the solvent to boil and rise up within the vessel. Hot solvent then comes into contact with a water-cooled reflux

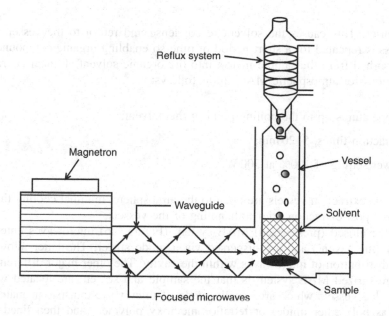

Figure 8.5 A schematic diagram of an atmospheric microwave-assisted extraction device. Figure drawn and provided by courtesy of Dr Pinpong Kongchana.

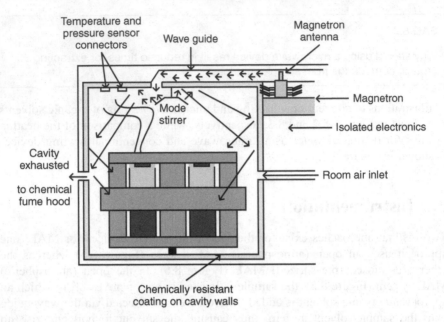

Figure 8.6 Schematic diagram of a pressurized microwave-assisted extraction device. Figure drawn and provided by courtesy of Dr Pinpong Kongchana.

condenser. This causes the solvent to condense and return to the vessel. This process is repeated for a short period of time so enabling organic compounds to be desorbed from the sample matrix into the organic solvent. Typical operating conditions for atmospheric MAE are as follows:

- temperatures up to the boiling point of the solvent;
- extraction times, 5–20 min;
- power setting of 100% at 300 W.

As the extraction vessels are open to the atmosphere, minimal cooling time is required post-extraction prior to handling of the vessels.

In the closed (pressurized) MAE system (Figure 8.6) microwaves enter the cavity (the 'oven') and are dispersed via a mode stirrer. The latter allows an even distribution of microwaves within the cavity. The other major difference in the pressurized MAE system is that the sample and solvent are located within the sealed vessels which are usually made of microwave-transparent materials, such as poly(ether imide) or tetrafloromethoxy polymers and then lined with

Teflon or perfluoroalkoxy polymers. In addition, at least one of the vessels has temperature/pressure controls which allow 'set conditions' to be used for extraction. Typical operating conditions for pressurized MAE are as follows:

- pressure, <200 psi;
- temperature within the range 110–145°C;
- extraction times, 5–20 min;
- power setting of 100% at 900 W.

As the extraction vessels are sealed, a cooling time of 20–30 min is applied post-extraction and prior to opening of the vessels.

A range of suppliers of commercial MAE systems is now available and some will now be briefly described.

8.2.1 Anton-Parr

The Multiwave 3000 system (www.anton-paar.com – last accessed on 4 January 2009) provides a flexible platform. It consists of two magnetrons capable of delivering power of up to 1400 W. This flexible system allows for extraction of 8, 16 or 48 samples by replacing the sample carrying rotor. The 8-vessel rotor allows continuous pressure monitoring within all of the 40–50 ml volume PTFE–TFM vessels whereas the 16-vessel rotor uses 50 ml volume PTFE–TFM vessels. High extraction throughput can be achieved with the 48-vessel rotor on 25 ml volume PFA vessels. The maximum operating conditions within the vessels range from a pressure of 20 bar (290 psi) and 200°C for the 25 ml volume vessels to 80 bar (1160 psi) and 300°C for the 40–50 ml volume vessels. The 16 and 48 sample rotors include one reference vessel with a wireless-controlled immersing temperature probe and pressure sensor.

8.2.2 CEM Corporation

The MARS system (www.cem.com – last accessed on 4 January 2009) is available in two formats for extraction. System 1 allows for up to 40 samples to be extracted simultaneously while system 2 allows up to 14 extractions simultaneously with optical fibre determination of temperature and pressure. It is capable of operating at a power of up to 1600 W. The 40-vessel rotor allows 'contactless' all-vessel continuous temperature monitoring within all 10, 25, 55 or 75 ml volume TFM or PFA vessels. The maximum operating conditions within the vessels range from temperatures of up to 260°C for the PFA vessels to 300°C for the TFM vessels. Alternatively, the 14-vessel rotor extractions can be performed at temperatures up to 200°C or pressures up to 200 psi in 100 ml PFA, Teflon or glass-lined vessels.

8.2.3 Milestone

The Ethos EX extraction system (www.milestonesci.com – last accessed on 4 January 2009) has two magnetrons capable of delivering power of up to 1600 W. The flexible system allows for extraction of 6, 12 or 24 samples in TFM vessels by replacing the sample-carrying rotor. The temperature can be measured in one vessel by using a fibre optic probe (up to 300°C) or via a 'contactless' infrared temperature monitor in all vessels. The 6-vessel rotor is designed for large samples (up to 40 g) and has a volume of 270 ml and is capable of operating at pressures up to 10 bar (150 psig) and a maximum temperature of 170°C. In contrast, the 12-vessel system can handle samples of up to 20 g and has a volume of 100 ml. These vessels are capable of operating at pressures up to 35 bar (500 psig) with a maximum temperature of 260°C. Similarly, the 24-vessel system can handle samples of up to 20 g and has a volume of 100 ml. These vessels are capable of operating at pressures up to 30 bar (435 psig) with a maximum temperature of 250°C. In addition, the WerTEX™ system provides an integrated approach for addition, filtration, evaporation and solvent recovery. Also, stirring can be achieved in all vessels by using an independently rotated magnet so allowing homogenous temperature distribution within each extraction vessel; stir bars are available in PTFE, 'Weflon', glass or quartz.

In all cases, a final stage is always required to separate the organic compound-containing solvent from the sample matrix. This is normally affected by filtering and/or centrifugation. The extract may be further pre-concentrated using solvent–evaporation approaches (see Chapter 1, Section 1.5.3).

8.3 Applications of MAE

An important aspect in using MAE for the recovery of organic compounds from sample matrices is whether the use of microwave energy has any influence on the stability of the compounds investigated. Liazid *et al.* [5] have investigated the influence of MAE on the stability of 22 phenols including benzoic acids, benzoic aldehydes, cinnamic acids, catechins, coumarins, stilbens and flavanols (Figure 8.7). In each case, 1 ml of the phenol was placed in 20 ml of methanol and subjected to a power of 500 W over a temperature of 50–175°C for 20 min (using an 'ETHOS-1600', Milestone system). It was observed (Table 8.2) that:

- Temperatures up to 100°C for 20 min produce no significant phenol degradation.

- The fewer the substituents on the aromatic ring, the higher the MAE stability.

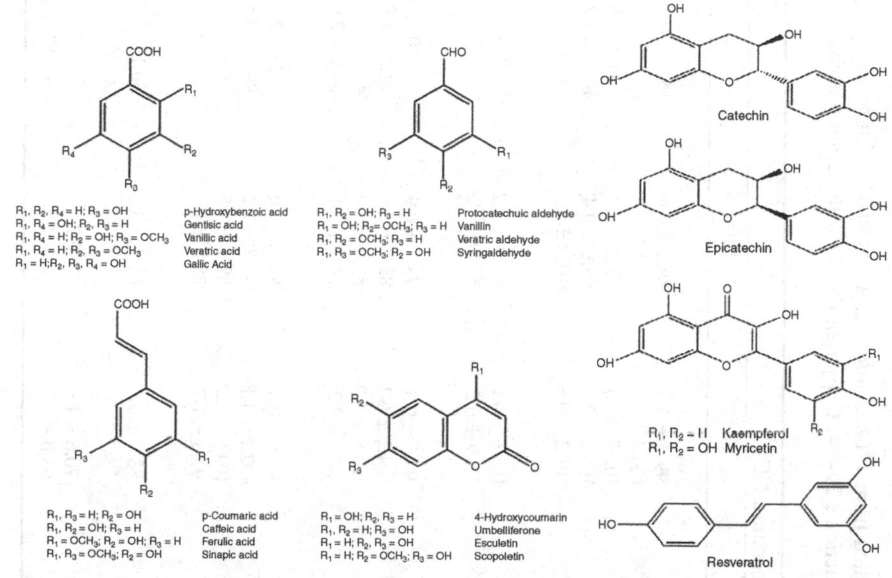

Figure 8.7 Phenol stability under MAE conditions: compounds investigated [5]. Reprinted from *J. Chromatogra., A.*, **1140**(1/2), Liazid *et al.*, 'Investigation on phenolic compounds stability during microwave-assisted extraction', 29–34, Copyright (2007) with permission from Elsevier.

- When two compounds have an equal number of substituents in the ring, the hydroxylates will be more easily degradable than the methoxylates.

A review of recent applications of MAE for organic compounds in analytical sciences is shown in Table 8.3.

DQ 8.2

What are the advantages and disadvantages of using MAE for recovery of organic compounds from sample matrices?

Answer

MAE has the main advantages of being able to extract multiple samples simultaneously using minimal organic solvent. Its main disadvantage is the relatively high capital cost and maintenance of the system for effective operation.

Microwave-assisted extraction has been applied to a diverse range of sample types (soils, sediments, sewage sludge, plants, marine samples) for the

Table 8.2 Results of phenol stability testing under MAE conditions [5]. Reprinted from *J. Chromatogra., A.,* **1140**(1/2), Liazid *et al.,* 'Investigation on phenolic compounds stability during microwave-assisted extraction', 29–34, Copyright (2007) with permission from Elsevier

Compound	Temperature (°C)					
	50	75	100	125	150	175
Benzoic acids						
Gentisic acid	97.1 ± 6.9	102.8 ± 3.9	103.5 ± 1.7	99.4 ± 3.3	89.6 ± 0.8	15.0 ± 16.6
Gallic acid	96.0 ± 6.4	102.3 ± 4.4	104.0 ± 1.3	100.4 ± 4.5	94.5 ± 0.7	17.3 ± 24.6
p-Hydroxybenzoic acid	97.1 ± 5.6	102.8 ± 3.1	104.6 ± 2.0	102.7 ± 3.1	95.7 ± 0.2	94.6 ± 1.8
Vanillic acid	96.5 ± 5.3	102.4 ± 4.5	102.9 ± 0.7	100.8 ± 3.4	93.9 ± 0.4	81.6 ± 4.7
Veratric acid	99.4 ± 6.4	111.1 ± 8.0	103.8 ± 2.9	101.2 ± 0.8	101.2 ± 1.8	101.2 ± 3.1
Flavan-3-ols						
(+)-Catechin	100.8 ± 6.3	103.1 ± 5.4	99.2 ± 1.4	105.6 ± 9.0	122.1 ± 4.2	0.0
(−)-Epicatechin	100.9 ± 2.7	96.7 ± 0.8	99.5 ± 2.3	75.9 ± 6.2	52.5 ± 5.2	0.0
Benzoic aldehydes						
Protocatechuic aldehyde	98 ± 3.6	102.7 ± 1.3	101.1 ± 6.6	99.3 ± 9.3	78.9 ± 0.9	0.0
Vanillin	97.0 ± 6.1	101.6 ± 4.9	95.5 ± 6.3	104.2 ± 6.1	82.4 ± 2.8	74.2 ± 1.8
Syringaldehyde	100.0 ± 5.4	101.7 ± 3.3	97.2 ± 6.8	103.8 ± 5.9	88.6 ± 1.0	47.5 ± 9.6
Veratric aldehyde	99.0 ± 3.4	100.1 ± 2.2	97.3 ± 9.1	94.7 ± 3.3	81.7 ± 3.1	84.0 ± 2.6
Cinnamic acids						
Caffeic acid	97.9 ± 10.2	96.2 ± 3.8	95.7 ± 1.2	96.9 ± 0.4	83.7 ± 3.0	1.9 ± 42.1
p-Coumaric acid	98.6 ± 7.4	97.2 ± 2.5	99.5 ± 2.4	102.2 ± 1.8	92.7 ± 1.2	99.4 ± 5.5
Ferulic acid	96.9 ± 7.7	93.4 ± 0.7	100.8 ± 0.8	104.9 ± 2.2	89.1 ± 1.5	0.0
Sinapic acid	96.8 ± 7.3	89.0 ± 1.6	98.5 ± 6.4	97.6 ± 1.7	78.8 ± 1.0	0.0
Estilben						
Resveratrol	102.9 ± 4.0	94.4 ± 1.8	98.8 ± 2.7	83.0 ± 6.1	61.2 ± 3.7	30.1 ± 5.8
Flavonols						
Myricetin	96.1 ± 3.0	96.7 ± 0.4	91.9 ± 1.5	70.6 ± 1.6	0.0	0.0
Kaempferol	100.2 ± 1.8	97.6 ± 1.0	99.3 ± 0.5	95.0 ± 3.0	0.0	0.0

Table 8.3 Selected examples of the use of microwave-assisted extraction (MAE) in analytical sciences[a]

Compounds	Matrix	Typical recoveries	Comments	Reference
MAE of compounds from soil matrices				
Chlorinated pesticides including mirex, α- and β-chlordane, *p*,*p*′-DDT, heptachlor, heptachlor epoxide, γ-hexachlorocyclohexane, dieldrin, endrin, aldrin and hexachlorobenzene	Soil	Recoveries of 8–51% were obtained with precision of 14–36% RSD	HS–SPME method optimized. Process led to reduced use of organic solvent and no requirement for extract clean-up. Detection limits ranged from 0.02 to 3.6 ng/g. Analysis by HS–SPME–GC–MS	6
Organophosphorus pesticides (OPPs) including diazinon, parathion, methyl pirimiphos, methyl parathion, ethoprophos and fenitrothion	Soil	Recoveries >73%, except for methyl parathion in some soils; precision <11% RSD	MAE took place with a water–methanol mixture for desorption and simultaneous partitioning on *n*-hexane. Addition of KH_2PO_4 to extraction solution increased recoveries. Analysis by GC–FPD gave detection limits in the range 0.004 to 0.012 µg/g	7
Polycyclic aromatic hydrocarbons (PAHs)	Soil	Recoveries ranged from 60 to 100%	Simultaneous MAE using *n*-hexane and hydrolysis of samples with methanolic potassium hydroxide. Sample extracts were cleaned-up with florisil and silica SPE cartridges connected in series prior to analysis by HPLC–DAD/Fl	8

(*Continued overleaf*)

Table 8.3 (*continued*)

Compounds	Matrix	Typical recoveries	Comments	Reference
MAE of compounds from sediment matrices				
Polycyclic aromatic hydrocarbons (PAHs), polychlorinated biphenyls (PCBs), phthalate esters (PEs), nonylphenols (NPs) and nonylphenol mono- and di-ethoxylates	Sediment	Good recoveries obtained: PAHs and PCBs validated with recoveries from CRM NIST 1944	p-MAE was carried out at 21 psi, 80% power with 15 ml of acetone. Filtered extracts were then fractionated using a florisil SPE cartridges: PAHs and PCBs eluted with *n*-hexane/toluene (4:1, vol/vol) and PEs, NPs and ethoxylates eluted with ethyl acetate. Analysis by GC–MS	9
Polybrominated diphenyl ethers (PBDEs), polybrominated biphenyls (PCBs) and polychlorinated naphthalenes (PCNs)	Sediment	Recoveries ranged from 75 to 95% with RSDs of 4–13%	MAE carried out using 48 ml of hexane/acetone (1:1, vol/vol) at 152°C and an extraction time of 24 min on 5 g of sample. Detection limits ranged from 4 to 20 pg/g, dry weight. Analysis by GC–MS	10
Triclosan and possible transformation products: 2,4-dichlorophenol and 2,4,6-trichlorophenol	Sediment	Recoveries ranged from 78 to 106%	MAE carried out using methanol/acetone (1:1, vol/vol); extract was centrifuged and diluted with NaOH and extracted with *n*-hexane. After concentration, sample extracts were silylated prior to analysis. Quantification limits ranged from 0.4 to 0.8 ng/g. Analysis by GC–MS–MS	11

Short-chain chlorinated alkanes	Sediment (river)	Recoveries were >90%; precision 7%	MAE carried out using 30 ml of n-hexane/acetone (1:1, vol/vol) at 115°C and an extraction time of 15 min on 5 g of sample. The detection limits was 1.5 ng/g. Analysis by GC–MS	12

MAE of compounds from sewage sludge matrices

Polycyclic aromatic hydrocarbons (PAHs)	Sewage sludge	Recoveries from CRM 088 ranged from 52 to 110%	Microwave procedure optimized for microwave power, irradiation time and extractant volume. Detection limits were between 4 and 12 ng/g. Analysis by HPLC–DAD/Fl	13
Nonylphenol (NP) and nonylphenol ethoxylates (NPEO)	Sewage sludge	Recoveries ranged from 61.4 (NPEO) to 91.4% (NP) with RSD <5%	Detection limits were 1.82 µg/g for NPEO and 2.86 µg/g for NP. Results compared with Soxhlet extraction and sonication. Analysis by HPLC	14
Polybrominated diphenyl ethers (PBDEs)	Sewage sludge	Recoveries ranged from 80 to 110%	MAE carried out using n-hexane/acetone (1:1, vol/vol) at 130°C and an extraction time of 35 min. Analysis by GC–MS	15

(Continued overleaf)

Table 8.3 (*continued*)

Compounds	Matrix	Typical recoveries	Comments	Reference
MAE of compounds from miscellaneous matrices				
Chlorophenols (17)	Incinerator ash	Recoveries ranged from 72 to 94%	Simultaneous derivatization with acetic anhydride in the presence of triethylamine (TEA) and extraction with a mixture of *n*-hexane and acetone was carried out using p-MAE. Optimization parameters considered were: volume of TEA and acetic anhydride, extraction time, temperature and volume of extraction solvent. Quantification limits were 2 to 5 ng/g using GC–MS	16
Organochlorine pesticides (21)	Vegetation (plants)	Recoveries ranged from 81.5 to 108.4%	Samples extracted using *n*-hexane/acetone (1:1, vol/vol) followed by extract clean-up with florisil and aminina SPE cartridges. Pesticides eluted with *n*-hexane/ethyl acetate (80:20, vol/vol) and analysed using GC–ECD. Method compared to Soxhlet extraction with similar results	17

Organochlorine pesticides (16)	Sesame seeds	Recoveries >80%; precision <12%	Samples extracted using water/acetonitrile followed by extract clean-up with florisil SPE cartridges. Optimization parameters considered were: extraction solvent, temperature, time and extractant volume. Quantification limits were in the range 5–10 µg/g using GC–MS	18
trans-Resveratrol	*Rhizma Polygoni Cuspidati* (Chinese medicinal herb)	Recoveries 93.7–103.2%; precision <3%	Samples extracted using 1-*n*-butyl-3-methylimidazolium-based ion liquid aqueous solutions as extraction solvent; specifically, 1-butyl-3-methylimidazolium bromide. Optimization parameters considered were: size of sample, liquid/solid ratio, extraction temperature and time	19
Polybrominated biphenyls (PBBs) and polybrominated diphenyl ethers (PBDEs)	Aquaculture feed samples	Acceptable accuracy obtained with respect to CRM values; precision <15%	Samples extracted using 14 ml of hexane/dichloromethane (1:1, vol/vol) for 15 min at 85°C. Method validated on IAEA-406 and WMF-01. Detection limits ranged from 10 to 600 pg/g. Extracts analysed using HS–SPME–GC–MS/MS	20

[a] Analytical techniques: HPLC–FL, high performance liquid chromatography with fluorescence detection; HPLC–DAD, high performance liquid chromatography with diode array detection; GC–ECD, gas chromatography with electron capture detection; GC–MS, gas chromatography–mass spectrometry; GC–FPD, gas chromatography with flame photometric detection; HS–SPME–GC, headspace–solid phase microextraction coupled with gas chromatography.

determination of organic compounds. All of the applications described (Table 8.3) use pressurized MAE, probably due to its commercial availability. It is possible to suggest some recommendations for the utilization of pressurized MAE in the extraction of organic compounds from samples, as follows.

- *Temperature*: >115°C but <145°C.

- *Pressure*: Operating at <200 psi.

- *Microwave power*: 100%.

- *Extraction time ('time at parameter')*: >5 min but no need to extend beyond 20 min. The longer time is recommended when >12 vessels are to be extracted simultaneously.

- *Extraction solvent volume*: 30–45 ml per 2–5 g of sample within a 100 ml volume extraction vessel.

- *Extraction solvent*: hexane/acetone (1:1, vol/vol) is commonly used; other solvents also appear useful, including ionic liquids.

SAQ 8.3

It is an important transferable skill to be able to search scientific material of importance to your studies/research. Using your University's Library search engine search the following databases for information relating to the extraction techniques described in this chapter, i.e. microwave-assisted extraction. Remember that often these databases are 'password-protected' and require authorization to access. Possible databases include the following:

- Science Direct;

- Web of Knowledge;

- The Royal Society of Chemistry.

(While the use of 'google' will locate some useful information please use the above databases.)

Summary

This chapter describes an important extraction technique for recovering organic compounds from solid samples, i.e. microwave-assisted extraction. The variables in selecting the most effective approach for microwave-assisted extraction are described. The commercial instrumentation for microwave-assisted extraction is also described. A review of applications of microwave-assisted extraction high-lights the diversity of application of this technique.

References

1. Abu-Samra, A., Morris, J. S. and Koirtyohann, S. R., *Anal. Chem.*, **47**, 1475 (1975).
2. Ganzler, K., Salgo, A. and Valko, K., *J. Chromatogr., A*, **371**, 299 (1986).
3. Papoutsis, D., *Photon. Spect.*, 53 (March, 1984).
4. Hasty, E. and Revesz, R., *Am. Lab.*, 66 (February, 1995).
5. Liazid, A., Palma, M., Brigui, J. and Barroso, C. G., *J. Chromatogr., A*, **1140**, 29–34 (2007).
6. Herbert, P., Morais, S., Paiga, P., Alves, A. and Santos, L., *Anal. Bioanal. Chem.*, **384**, 810–816 (2006).
7. Fuentes, E., Baez, M. A. and Labra, R., *J. Chromatogr., A*, **1169**, 40–46 (2007).
8. Pena, M. T., Pensado, L., Casais, M. C., Mejuto, M. C. and Cela, R., *Anal. Bioanal. Chem.*, **387**, 2559–2567 (2007).
9. Bartolome, L., Cortazar, E., Raposo, J. C., Usobiaga, A., Zuloaga, O., Etxebarria, N. and Fernandez, L. A., *J. Chromatogr., A*, **1068**, 229–236 (2005).
10. Yusa, V., Pardo, O., Pastor, A. and de la Guardia, M., *Anal. Chim. Acta*, **557**, 304–313 (2006).
11. Morales, S., Canosa, P., Rodriguez, I., Rubi, E. and Cela, R., *J. Chromatogr., A*, **1082**, 128–135 (2005).
12. Parera, J., Santos, F. J. and Galceran, M. T., *J. Chromatogr., A*, **1046**, 19–26 (2004).
13. Villar, P., Callejon, M., Alonso, E., Jimenez, J. C. and Guiraum, A., *Anal. Chim. Acta*, **524**, 295–304 (2004).
14. Fountoulakis, M., Drillia, P., Pakou, C., Kampioti, A., Stamatelatou, K. and Lyberatos, G., *J. Chromatogr., A*, **1089**, 45–51 (2005).
15. Shin, M., Svoboda, M. L. and Falletta, P., *Anal. Bioanal. Chem.*, **387**, 2923–2929 (2007).
16. Criado, M. R., da Torre, S. P., Pereiro, I. R. and Torrijos, R. C., *J. Chromatogr., A*, **1024**, 155–163 (2004).
17. Barriada-Pereira, M., Concha-Grana, E., Gonzalez-Castro, M. J., Muniategui-Lorenzo, S., Lopez-Mahia, P., Prada-Rodriguez, D. and Fernandez-Fernandez, E., *J. Chromatogr., A*, **1008**, 115–122 (2003).
18. Papadakis, E. N., Vryzas, Z. and Papadopoulou-Mourkidou, E., *J. Chromatogr., A*, **1127**, 6–11 (2006).
19. Dou, F.-Y., Xiao, X.-H. and Li, G.-K., *J. Chromatogr., A*, **1140**, 56–62 (2007).
20. Carro, A. M., Lorenzo, R. A., Fernandez, F., Phan-Tan-Luu, R. and Cela, R., *Anal. Bioanal. Chem.*, **388**, 1021–1029 (2007).

Chapter 9
Matrix Solid Phase Dispersion

9.1 Introduction

Matrix solid phase dispersion (MSPD) is used for the extraction and fractionation of solid, semi-solid or viscous biological samples. The process of MSPD is analogous to solid phase extraction (SPE), as described in Chapter 4. Recently, several reviews have appeared that summarize developments in the use of MSPD in drug, tissue and food analysis [1, 2]. The concept of MSPD is that a sample is mixed with a support material, e.g. octadecylsilane (C18), alumina or 'florisil' in a glass or agate mortar and 'pestle' for approximately 30 s.

DQ 9.1

What will be the effect of this mechanical grinding on the sample?

Extraction Techniques in Analytical Sciences   John R. Dean
© 2009 John Wiley & Sons, Ltd

Answer

The mechanical grinding of the sample with the support acts as an abrasive, leading to shearing and disruption of the sample matrix, so producing a large surface area for solvent interaction.

The blended sample mixture is then quantitatively transferred to a column fitted with a frit (e.g. an empty SPE cartridge). By addition of single or multiple solvents, it is then possible to perform clean-up and or (selective) elution of compounds (Figure 9.1).

Important factors in MSPD include the following:

- Particle size of support material: 40–100 µm is an ideal compromise between restricted flow that can result from the use of smaller particle sizes (3–10 µm) and cost of the support.

- Use of end-capped or non-end-capped support materials, e.g. ODS, with different carbon loadings (i.e. 10–20%).

- Use of other support materials e.g. alumina, 'florisil' or silica.

- Ratio of sample to support material. The ratio of sample to sorbent varies between 1:1 and 1:4 wt/wt, e.g. 0.5 g of sample to 2.0 g of C18 (1:4 wt/wt).

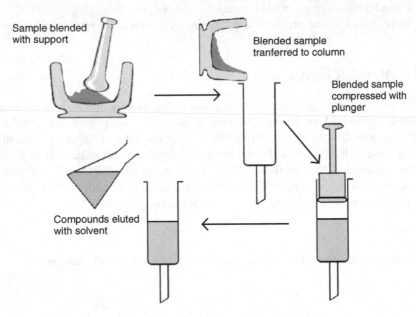

Figure 9.1 Schematic diagram of matrix solid phase dispersion.

- Addition of chelating agents, acids and bases may affect clean-up and elution of compound(s).

- Selection of solvent(s) for clean-up, i.e. removal of extraneous material, e.g. fats.

- Selection of solvent(s) for elution of compound(s).

- Elution volume, i.e. for a 0.5 g sample mixed with 2.0 g of support material then the target compounds typically elute in the first 4 ml of solvent.

- Influence of the sample matrix itself, i.e. the different properties of the sample will influence the recovery of target compounds.

- Whether additional clean-up procedures, e.g. alumina SPE, are required prior to instrumental analysis.

SAQ 9.1

Where would you find the use of C18 material of 40–100 μm particle size?

SAQ 9.2

Where would you find the use of C18 material of 3–10 μm particle size?

SAQ 9.3

What does the process of end-capping do to a C18 sorbent phase?

A typical procedure for performing matrix solid phase dispersion extraction is shown in Figure 9.2.

9.2 Issues on the Comparison of MSPD and SPE

While MSPD is similar in appearance to solid phase extraction (Chapter 4) its performance and function are different. MSPD differs primarily in the following respects:

(1) The sample is dissipated, by mixing with the support material over a large surface area (no similar process takes place in SPE).

(2) The sample is homogeneously distributed through the column (in SPE the sample is loaded on top of the sorbent).

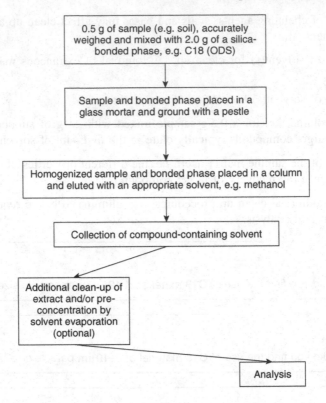

Figure 9.2 Typical procedure for matrix solid phase dispersion.

9.3 A Review of Selected Applications

A range of applications using MSPD are reviewed in Table 9.1. This approach has been applied to a diverse range of sample types ranging from liquid samples (e.g. fruit juices) to (semi)-solid samples in the form of biological tissues (e.g. fish tissue), plant materials (cereals) and food matrices (e.g. potato chips). A range of sorbents have been used including C18, 'florisil', alumina, aminopropyl and silica gel, producing good recoveries (ranging from 61–116%) with typical RSDs < 12%.

DQ 9.2

What other sample types are there to which you might apply MSPD?

Answer

Other sample types might include soil, sediment and sewage sludge, as well as fruits and vegetables.

Table 9.1 Selected examples of the use of matrix solid phase dispersion (MSPD) in analytical sciences[a]

Compounds	Matrix	Typical recoveries	Comments	Reference
Endosulphan isomers and endosulphan sulfate	Tomato juice	Recoveries ranged from 81 to 100% with RSD < 10%	Parameters optimized were: type of adsorbent, extraction solvent and extraction assistance using sonication. Detection limits were 1 µg/kg. Method applied to commercial samples; some found to contain compounds between 1 and 5 µg/kg. Analysis by GC–ECD	3
Herbicides (15)	Fruit juices (carrot, grape and multivegetable juices)	Recoveries ranged from 82 to 115% with RSD < 10%	Method used 'florisil' packed in glass columns and subsequent extraction with ethyl acetate with assisted extraction using sonication. Detection limits ranged from 0.1 to 1.6 µg/l. Method applied to commercial juice samples. Analysis by GC–MS	4
Aflatoxin B1, B2, G1 and G2	Peanuts	Recoveries ranged from 78 to 86% with RSD 4–7%	Parameters optimized were: type of solid support and elution solvent. Method used 2 g of sample, 2 g of C18 bonded silica (as MSPD sorbent) and acetonitrile as eluting solvent. Quantitation limits ranged from 0.125 to 2.5 ng/g. Analysis by HPLC–Fl	5

(Continued overleaf)

Table 9.1 (*continued*)

Compounds	Matrix	Typical recoveries	Comments	Reference
Carbendazim	Plant material (cereal samples)	Recoveries ranged from 84.3 to 90.7% with RSD 2.7–4.1% at fortification levels of 0.04, 0.08 and 0.1 µg/g	On-line coupling of MSPD with HPLC compared to off-line approach. Method of standard additions used for quantitation. Detection limit was 0.02 µg/g. Analysis by HPLC–UV	6
Ethylene bisdithiocarbamates main metabolites (ethylenethiourea and ethylenebis(isothiocyanate) sulfide).	Plant material (almond samples)	Recoveries ranged from 76 to 85% with RSD 3–12%	Method used 0.2 g of sample, washed sand (as MSPD sorbent) and NaOH as de-fatting agent. Extracts cleaned-up using an alumina cartridge with an eluting solvent of acetonitrile. Quantitation limits ranged from 0.05 to 0.07 mg/kg. Analysis by HPLC–DAD	7
Pesticides (malathion, methyl parathion and β-endosulphan)	Rice	Recoveries ranged from 75.5 to 116% with RSD 0.5–10.9%	Parameters optimized were: sample and solid support amounts, adsorbent and eluting solvent. Detection limits ranged from 20 to 105 pg. Method applied to commercial rice samples. Analysis by GC–MS	8

Pesticides (OCPs and pyrethroids)	Tea samples	Recoveries > 80% and precision < 7% for fortification levels in the range 0.01–0.05 mg/kg	Parameters optimized were: sorbent type, eluent composition, dichloromethane concentration and eluting volume. Method used 'florisil' as sorbent and n-hexane/dichloromethane (1:1, vol/vol) as eluent. Quantification limits were in the range 0.002–0.06 mg/kg. Analysis by GC	9
Acrylamide	Potato chips	Good recoveries	Samples were ground (0.5 g) and dispersed in 2 g C18 before being placed in an empty column; following a clean-up with n-hexane (removes fat), the compound was eluted with water (4 ml + 4 ml). Extracts were brominated prior to analysis by GC–MS. Quantification and detection limits were 38.8 and 12.8 µg/kg, respectively	10
Pesticides	Olives and olive oil	Recoveries 85 and 115% and precision < 10% for a range of fortification levels	Samples were dispersed with aminopropyl as sorbent followed by clean-up, in the elution step, with 'florisil'. Olive oil samples were pre-treated using LLE. Extracts were analysed by either GC–MS or HPLC–MS. Detection limits were in the range 10 to 60 µg/kg by GC–MS and < 5 µg/kg by LC–MS	11

(Continued overleaf)

Table 9.1 (*continued*)

Compounds	Matrix	Typical recoveries	Comments	Reference
Polychlorinated biphenyls (PCBs)	Butter, chicken and beef fat	Quantification limits of 0.4 ng of each PCB per g of fat were achieved	Evaluation of different normal phase sorbents and elution solvents carried out with respect to extraction yield and lipids removal efficiency. Optimal conditions consisted of 0.5 g of sample dried with anhydrous sodium sulfate, dispersed on 1.5 g of 'florisil' and transferred to an SPE cartridge which already contained 5 g of 'florisil'. 'Non-coplanar' PCBs were eluted with 15 ml of *n*-hexane. 'Coplanar' and 'non-coplanar' PCBs eluted with 20 ml of hexane/dichloromethane (90:10, vol/vol). Extracts were evaporated to 0.2 ml and then analysed by either GC–MS or GC–ECD	12
Thyreostatic compounds, including 2-thiouracil, 6-methyl-2-thiouracil, 6-propyl-2-thiouracil, 6-pheyl-2-thiouracil and 1-methyl-2-mercpto-imidazole	Animal tissues	Recoveries > 70% and precision between 4.5 and 8.7% RSD	Samples were dispersed with silica gel (sorbent). Extracts were derivatized with pentafluorobenzylbromide in a strong basic medium and then with *N*-methyl-*N*-(trimethylsilyl)-trifluoroacetamide prior to analysis by GC–MS. Detection limits ranged from 10 to 50 µg/kg	13

Analyte	Matrix	Method	Recoveries	Ref.
Polycyclic aromatic hydrocarbons (PAHs)	Fish tissue	Samples (0.6–0.8 g) were dispersed with 2 g of C18 sorbent and 0.5 g anhydrous sodium sulfate and placed in an SPE cartridge pre-loaded with 2 g of 'florisil' and 1 g C18. Cartridges were eluted with acetonitrile prior to analysis by HPLC–Fl. Detection limits ranged from 0.04 to 0.32 ng/g	Recoveries > 80%	14
20 Organochlorine pesticides (OCPs) and 8 polychlorinated biphenyls (PCBs)	Chicken eggs	Samples were dispersed with 'florisil' (sorbent) and eluted with dichloromethane/hexane (1:1, vol/vol). Extracts were cleaned-up using concentrated sulfuric acid prior to analysis by GC–ECD. Detection limits were < 0.7 ng/g. Method used to analyse 30 commercial products	Recoveries 82–110% and precision < 8% RSD for samples fortified over the concentration range 10–200 ng/g	15
Imidacloprid, carbaryl and aldicarb (and their main metabolites)	Honeybees	Samples were dispersed with C18 (sorbent) and eluted with dichloromethane/methanol. Analysis by HPLC–APCI–MS. Detection limits ranged from 0.004 to 0.09 mg/kg. Method compared with an LLE approach	Recoveries ranged between 61 and 99% with precision < 14% RSD	16

[a] Analytical techniques: HPLC–UV, high performance liquid chromatography with ultraviolet detection; HPLC–Fl, high performance liquid chromatography with fluorescence detection; HPLC–DAD, high performance liquid chromatography with diode array detection; GC–ECD, gas chromatography with electron capture detection; GC–MS, gas chromatography–mass spectrometry; GC–FPD, gas chromatography with flame photometric detection; SPME–GC–MS, solid phase microextraction coupled with gas chromatography–mass spectrometry; HS–SPME–GC, headspace–solid phase microextraction coupled with gas chromatography.

SAQ 9.4

It is an important transferable skill to be able to search scientific material of importance to your studies/research. Using your University's Library search engine search the following databases for information relating to the extraction techniques described in this chapter, i.e. matrix solid phase dispersion. Remember that often these databases are 'password-protected' and require authorization to access. Possible databases include the following:

- Science Direct;

- Web of Knowledge;

- The Royal Society of Chemistry.

(While the use of 'google' will locate some useful information please use the above databases.)

Summary

A relatively new approach for recovering organic compounds from (semi)-solid samples, i.e. matrix solid phase dispersion, is described in this chapter. The important variables in selecting the most effective approach for matrix solid phase dispersion are described. A review of applications of matrix solid phase dispersion highlights the application of this technique.

References

1. Barker, S. B., *J. Biochem. Biophys. Meth.*, **70**, 151–162 (2007).
2. Kristenson, E. M., Ramos, L. and Brinkman, U. A. Th., *Trends Anal. Chem.*, **25**, 96–111 (2006).
3. Albero, B., Sanchez-Brunete, C. and Tadeo, J. L., *J. Chromatogr., A*, **1007**, 137–143 (2003).
4. Albero, B., Sanchez-Brunete, C., Donoso, A. and Tadeo, J. L., *J. Chromatogr., A*, **1043**, 127–133 (2004).
5. Blesa, J., Soriano, J. M., Molto, J. C., Marin, R. and Manes, J., *J. Chromatogr., A*, **1011**, 49–54 (2003).
6. Michel, M. and Buszewski, B., *J. Chromatogr., B*, **800**, 309–314 (2004).
7. Garcinuno, R. M., Ramos, L., Fernandez-Hernando, P. and Camara, C., *J. Chromatogr., A*, **1041**, 35–41 (2004).
8. Dorea, H. S. and Sobrinho, L. L., *J. Brazil. Chem. Soc.*, **15**, 690–694 (2004).
9. Hu, Y.-Y., Zheng, P., He, Y.-H. and Sheng, G.-P., *J. Chromatogr., A*, **1098**, 188–193 (2005).
10. Fernades, J. O. and Soares, C., *J.Chromatogr., A*, **1175**, 1–6 (2007).
11. Ferrer, C., Gomez, M. J., Garcia-Reyes, J. F., Ferrer, I., Thurman, E. M. and Fernandez-Alba, A. R., *J. Chromatogr., A*, **1069**, 183–194 (2005).

12. Criado, M. R., Fernandez, D. H., Pereiro, I. R. and Torrijos, R. C., *J. Chromatogr., A*, **1056**, 187–194 (2004).
13. Zhang, L., Liu, Y., Xie, M.-X. and Qiu, Y.-M., *J. Chromatogr., A*, **1074**, 1–7 (2005).
14. Pensado, L., Casais, M. C., Mejuto, M. C. and Cela, R., *J. Chromatogr., A*, **1077**, 103–109 (2005).
15. Valsamaki, V. I., Boti, V. I., Sakkas, V. A. and Albanis, T. A., *Anal. Chim. Acta*, **573–574**, 195–201 (2006).
16. Totti, S., Fernandez, M., Ghini, S., Pico, Y., Fini, F., Manes, J. and Girotti, S., *Talanta*, **69**, 724–729 (2006).

Chapter 10

Supercritical Fluid Extraction

Learning Objectives

- To be aware of approaches for performing supercritical fluid extraction of organic compounds from solid samples.
- To understand the theoretical basis for supercritical fluid extraction.
- To understand the practical aspects of supercritical fluid extraction.
- To appreciate the potential variables when performing supercritical fluid extraction.
- To be aware of the practical applications of supercritical fluid extraction.

10.1 Introduction

A supercritical fluid is a substance which is above its critical temperature and pressure. The discovery of the supercritical phase is attributed to Baron Cagniard de la Tour in 1822 [1]. This can be explained by consideration of a phase diagram for a pure substance (Figure 10.1).

SAQ 10.1

What is a phase diagram?

For example, the solid–gas boundary corresponds to sublimation, the solid–liquid boundary corresponds to melting and the liquid–gas boundary corresponds to vaporization. The three curves intersect where the three phases

Extraction Techniques in Analytical Sciences   John R. Dean
© 2009 John Wiley & Sons, Ltd

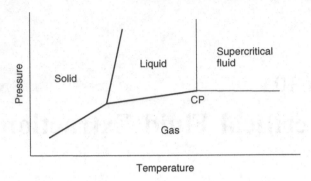

Figure 10.1 Schematic phase diagram for a pure substance. From Dean, J. R., *Extraction Methods for Environmental Analysis*, Copyright 1998. © John Wiley & Sons, Limited. Reproduced with permission.

co-exist in equilibrium, known as the triple point. At the critical point, designated by both a critical temperature and a critical pressure, no liquefaction will take place on raising the pressure and no gas will be formed on increasing the temperature – it is this defined region, which is by definition, the supercritical region. The use of supercritical fluids for extraction in analytical sciences was first developed in the mid-1980s [2]. A range of substances have been used for supercritical fluid extraction (SFE) (Table 10.1). The most common supercritical fluid in analytical sciences is carbon dioxide.

DQ 10.1

What advantages does CO_2 have as a supercritical fluid?

Table 10.1 Critical properties of selected substances

Substance	Critical temperature (°C)	Critical pressure	
		(atm)	(psi)
Ammonia	132.4	115.0	1646.2
Carbon dioxide	31.1	74.8	1070.4
Chlorodifluoromethane	96.3	50.3	720.8
Ethane	32.4	49.5	707.8
Methanol	240.1	82.0	1173.4
Nitrous oxide	36.6	73.4	1050.1
Water	374.4	224.1	3208.2
Xenon	16.7	59.2	847.0

Answer

It has the following properties:

- Moderate critical pressure (73.8 bar).
- Low critical temperature (31.1°C).
- Low toxicity and reactivity.
- High purity at low cost.
- Use for extractions at temperatures < 150°C.
- Ideal for extraction of thermally labile compounds.
- Ideal extractant for non-polar species, e.g. alkanes.
- Reasonably good extractant for moderately polar species, e.g. PAHs and PCBs.
- Can directly vent to the atmosphere.
- Little opportunity for chemical change in the absence of light and air.
- Being a gas at room temperature allows for direct coupling to GC and SFC equipment.

The major disadvantage of CO_2 is its non-polar nature (it has no permanent dipole moment) meaning that for a high proportion of applications its solvent strength is inadequate. This issue can be addressed by the addition of a polar organic solvent or 'modifier' to the supercritical fluid.

DQ 10.2

How might a modifier be added to the SFE system?

Answer

Addition of the modifier is possible in several ways including:

- Spiking of organic solvent directly to the sample in the extraction cell.
- Purchase of pre-mixed cylinders, e.g. 10% methanol-modified CO_2.
- Addition of a second pump that allows in-line mixing of CO_2 and organic solvent prior to the extraction vessel.

The major advantage of SFE is the diversity of properties that it can exhibit. These include:

- Variable solvating power (provides properties intermediate between gases and liquids).

- High diffusivity (allows penetration of solid matrices and mass transfer).

- Low viscosity (provides good flow characteristics and mass transfer).

- Minimal surface tension (allows the supercritical fluid to penetrate within low-porosity matrices).

These properties of a supercritical fluid allow selective extraction of organic compounds from sample matrices.

10.2 Instrumentation for SFE

The major components of an SFE system are as follows:

- a supply of high-purity carbon dioxide;

- a supply of high-purity organic modifier;

- two pumps;

- an oven;

- an extraction vessel;

- a pressure outlet or restrictor;

- a suitable collection vessel for quantitative recovery of extracted organic compounds.

DQ 10.3

Draw a schematic diagram of an SFE system based on the above description.

Answer

A schematic diagram of an SFE system is shown in Figure 10.2.

The choice of CO_2 is an important initial consideration as far as impurities are concerned. It is essential that the level of impurities encountered in the CO_2 do not interfere with the subsequent analysis. The CO_2 is supplied in a cylinder fitted with a dip tube which allows liquified CO_2 to be pumped by a reciprocating

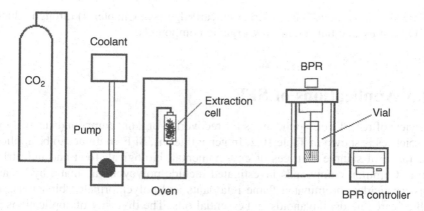

Figure 10.2 Schematic diagram of an SFE system.

or syringe pump. (NOTE: It is possible to purchase cylinders that contain both CO_2 and organic modifier, e.g. 10% methanol-modified CO_2). To allow pumping of the liquefied CO_2, without cavitation, requires the pump head to be cooled. This is achieved by using a jacketed pump head which is either cooled via an ethylene glycol mixture pumped using a re-circulating water bath or a 'peltier' device. If the modifier is to be added via a second pump (which does not require any pump head cooling) the CO_2 and modifier are mixed using a T-piece.

To achieve the required critical temperature requires the extraction vessel containing the sample to be located in an oven which is capable of effective controlled heating in the range 30–250°C. The sample vessel, made of stainless steel, must be capable of withstanding high pressures (up to 10 000 psi) safely. The sample is located inside the extraction vessel and often requires some pre-treatment and/or mixing with additional components to ensure effective extraction. For additional information, please see Chapter 7 on *Pressurized Fluid Extraction*, Section 7.4 (Method Development for PFE).

Pressure is established within the extraction vessel by using a variable (mechanical or electronically controlled) restrictor. The variable restrictor allows a constant, operator-selected flow rate whose pre-selected pressure is maintained by the size of the variable orifice. As a result of adiabatic expansion of the CO_2 upon exiting the restrictor the build up of ice is common unless the restrictor is heated. Sample extracts are collected in a vial prior to subsequent analysis as follows:

- In an open vial containing organic solvent.

- In a sealed vial containing solvent but with the addition of a solid phase extraction cartridge (see Chapter 4) through which CO_2 can escape but retains any organic compounds.

- Directly onto a solid phase extraction cartridge (see Chapter 4) through which CO_2 can escape but retains any organic compounds.

10.3 Applications of SFE

A review of recent SFE applications for recovering organic compounds in analytical sciences is shown in Table 10.2. In general terms, SFE continues to be applied to a range of sample matrices of environmental, biological, food and industrial origin. Common compounds investigated include polycyclic aromatic hydrocarbons, pesticides, brominated flame retardants and polychlorinated biphenyls, as well as carotenoids, flavanoids and essential oils. The diversity of applications is reflected in the use of a technology that uses a minimum of organic solvent and so would be labelled as 'environmentally friendly'. On that basis SFE is being used to extract natural products from medical plants [13] and essential oils from plants [8], as well as for monitoring risk to humans, e.g. PCBs in seaweed [11] and PAHs in vegetable oil [12].

10.4 Selection of SFE Operating Parameters

Important considerations for the selection of SFE operating conditions are as follows [13]:

- *Extraction temperature*
    - For thermolabile compounds the temperature should be within the range 35 to 60°C, i.e. close to the critical point but not so high a temperature that compound degradation might occur.
    - For non-thermally labile compounds the temperature can exceed 60°C (up to 200°C).
- *Extraction pressure*
    - The higher the pressure, the larger is the solvating power (often described in terms of CO_2 density which can vary between 0.15 and 1.0 g/ml) and the smaller is the extraction selectivity.
- *Flow rate of liquid CO_2*
    - A typical flow rate of 1 ml/min is used.
- *Extraction time*
    - Often a compromise between obtaining a good recovery and the duration of the process. Typical extraction times may range from 30 to 60 min.

Table 10.2 Selected examples of the use of supercritical fluid extraction (SFE) in analytical sciences[a]

Compounds	Matrix	Typical recoveries	Comments	Reference
SFE of compounds from soil and sediment matrices				
Polycyclic aromatic hydrocarbons (PAHs)	Soil and sediment	> 90% from spiked soils	Influence of 5% (vol/vol) organic modifier (methanol, n-hexane and toluene) on the supercritical CO_2 of PAHs from the sample. Influence of temperature (50 and 80°C) and pressure (230 to 600 bar) on recovery evaluated. Analysis by GC–MS	3
Pesticides (including OCPs, OPPs, triazine and acetanilide herbicides)	Soil	80.4–106.5% (RSDs, 4.2–15.7%) in the sub-ppb level (0.1–3.7 μg/kg)	Experimental design approach applied to optimize SFE conditions. Sample from an intensive horticultural area analysed by GC–MS–MS	4
SFE of compounds from industrial products				
Organohalogenated pollutants (15), including brominated flame retardants	Aquaculture samples (fish feed and shellfish samples)	Good	*In situ* supercritical CO_2 extraction and clean-up (using aluminium oxide basic and acidic silica gels). SFE parameters screened, using a factorial design, were extraction temperature, pressure, static extraction time, dynamic extraction time and CO_2 flow rate. The two most important variable were then optimized, i.e. pressure (165 bar) and dynamic extraction time (27 min). Excellent linearity, detection (0.01–0.2 ng/g) and quantification limits (0.05–0.8 ng/g) were obtained using GC–MS/MS	5

(Continued overleaf)

Table 10.2 (*continued*)

Compounds	Matrix	Typical recoveries	Comments	Reference
Carotenoids (lycopene and β-carotene), tocopherols and sitosterols	Industrial tomato by-products	90.1% for lycopene using supercritical CO_2 at 460 bar and 80°C	Supercritical CO_2 extraction was optimized (pressure and temperature); influence of samples from different sources and the effect of storage (air-dried versus seep-frozen) were investigated. Extraction yields were dependent upon experimental conditions. Analysis by GC, HPLC and TLC densitometry.	6
SFE of compounds from plant matrices				
Flavanoids (3) (orotinin, orotinin-5-methyl ether and licoagrochalcone B)	*Patrinia villosa* (medicinal plant)	Preparative SFE yield (2.82%) producing a combined yield of all 3 compounds of 0.82 mg/g dry weight	Supercritical CO_2 extraction was optimized with respect to pressure, temperature, modifier and sample particle size (analytical-scale); extraction scaled-up (×100) using a preparative system under the optimized conditions of 25 MPa, 45°C, particle size of 40–60 mesh and 20% methanol-modified supercritical CO_2	7
Sage essential oil	*Salvia officinalis* L.	Extraction yield of oxygenated monoterpene manool was more than double that obtained using hydrodistillation	Supercritical CO_2 extraction as follows: 9–12.8 MPA, 25–50°C, sage feed, 3–4 g; CO_2 flow rate, 0.05–0.35 g/min; solvent-to-feed ratio, 16–21	8

Hydrocarbons	*Euphorbia macroclada*	Sequential extraction using 10% (vol/vol) methanol-modified supercritical CO_2 using a pressure of 400 atm and temperature of 50°C, followed by sonication in DCM for an additional 4 h. Comparison with Soxhlet extraction using DCM for 8 h. All extracts were fractionated using a silica-gel column prior to GC analysis	Yield was 5.8% (compared to 1.1% by Soxhlet)	9
OCPs (α-, β-, γ- and δ-benzene hexachloride (BHC), pentachloronitro-benzene and DDT and its metabolites)	Ginseng	Extraction using 10% (vol/vol) ethanol-modified supercritical CO_2 using a pressure of 300 atm and temperature of 60°C, followed by collection using a C18 trap with *n*-hexane as eluting solvent. Comparison with Soxhlet extraction. Analysis by GC–ECD with MS confirmation	SFE more efficient than Soxhlet	10
SFE of compounds from food products				
PCBs (12)	Seaweed	Comparison of SFE with Soxhlet extraction from algae samples. Analysis by GC–ECD. Method applied to three real seaweed samples (only PCB101 found)	Similar recoveries between SFE and Soxhlet extraction. Precision was better for Soxhlet (< 3.9%) compared to SFE (< 9.2%)	11
PAHs	Vegetable oil	SFE of PAHs in vegetable oil. Analysis by HPLC–FL. Detection and quantification limits were < 1.55 µg/kg oil and < 2.55 µg/kg oil, respectively	Method allows evaluation of edible oil safety as part of consumer protection	12

[a] Analytical techniques: HPLC, high performance liquid chromatography; HPLC–FL, high performance liquid chromatography with fluorescence detection; GC–ECD, gas chromatography with electron capture detection; GC–MS, gas chromatography–mass spectrometry; GC, gas chromatography; TLC, thin layer chromatography.

- *Sample matrix particle size*

  - The smaller the uniform particle size, the more likely that efficient extraction takes place; however, a very small sample particle size can lead to 'channelling' in the sample extraction cell (leading to poor CO_2 to analyte interaction and consequently poorer extraction efficiency). Sample particle sizes in the range 0.25 to 2.0 mm are often used.

- *Addition of a modifier*

  - The lack of a permanent dipole in CO_2 means that polar compounds will often have poor recoveries. This situation is often addressed by the addition of a polar organic solvent modifier, typically 5 or 10% methanol (or ethanol).

Recommended initial SFE operating conditions:

- Supercritical CO_2 will generally solvate 'GC-able' compounds under extraction conditions of pressure, 400 atm and a temperature of $50°C$.

- For fairly polar or compounds with high molecular masses the addition of an organic modifier (10% vol/vol methanol or ethanol) may be necessary with a subsequent increase in temperature to $70°C$.

- For ionic compounds the addition of an ion-pairing reagent may be beneficial.

SAQ 10.2

It is an important transferable skill to be able to search scientific material of importance to your studies/research. Using your University's Library search engine search the following databases for information relating to the extraction techniques described in this chapter, i.e. supercritical fluid extraction. Remember that often these databases are 'password-protected' and require authorization to access. Possible databases include the following:

- Science Direct;

- Web of Knowledge;

- The Royal Society of Chemistry.

(While the use of 'google' will locate some useful information please use the above databases.)

Summary

This chapter describes an extraction technique for recovering organic compounds from solid samples, i.e. supercritical fluid extraction. The variables in selecting the most effective approach for supercritical fluid extraction are described. A review of applications of supercritical fluid extraction highlights the usefulness of this technique.

References

1. de la Tour, C., *Ann. Chim. (Paris)*, **21**, 127–132 (1822).
2. Dean, J. R. (Ed), *Applications of Supercritical Fluids in Industrial Analysis*, Blackie Academic and Professional, Glasgow, UK (1993).
3. Librando, V., Hutzinger, O., Tringali, G. and Aresta, M., *Chemosphere*, **54**, 1189–1197 (2004).
4. Goncalves, C., Carvalho, J. J., Azenha, M. A. and Alpendurada, M. F., *J. Chromatogr., A*, **1110**, 6–14 (2006).
5. Rodil, R., Carro, A. M., Lorenzo, R. A. and Cela, R., *Chemosphere*, **67**, 1453–1462 (2007).
6. Vagi, E., Simandi, B., Vasarhelyine, K. P., Daood, H., Kery, A., Dolescchall, F. and Nagy, B., *J. Supercrit. Fluids*, **40**, 218–226 (2007).
7. Peng, J., Fan, G., Chai, Y. and Wu, Y., *J. Chromatogr., A*, **1102**, 44–50 (2006).
8. Aleksovski, S. A. and Sovova, H., *J. Supercrit. Fluids*, **40**, 239–245 (2007).
9. Ozcan, A. and Ozcan, A. S., *Talanta*, **64**, 491–495 (2004).
10. Quan, C., Li, S., Tian, S., Xu, H., Lin, A. and Gu, L., *J. Supercrit. Fluids*, **31**, 149–157 (2004).
11. Punin Crespo, M. O. and Lage Yusty, M. A., *Chemosphere*, **59**, 1407–1413 (2005).
12. Lage Yusty, M. A. and Cortizo Davina, J. J., *Food Control*, **16**, 59–64 (2005).
13. Reverchon, E. and DeMarco, I., *J. Supercrit. Fluids*, **38**, 146–166 (2006).

GASEOUS SAMPLES

Chapter 11

Air Sampling

Learning Objectives

- To be aware of approaches for recovering organic compounds from air samples.
- To appreciate the range of techniques available for air sampling and their limitations and benefits.
- To be aware of the distinction between active and passive sampling.
- To understand the theoretical aspects of passive sampling.
- To be aware of the applications of air sampling.

11.1 Introduction

The trace analysis of volatile organic compounds (VOCs) in the atmosphere, workplace and on industrial sites needs to be monitored with regard to safety considerations, e.g. emissions to the atmosphere or occupational standards. Typical VOCs determined in the atmosphere are shown in Table 11.1. In order to differentiate between individual compounds it is necessary to use gas chromatography (GC) with either a flame ionization detector (FID), electron capture detector (ECD) or mass spectrometer (MS) (see Chapter 1). The low concentration of VOCs in air often means that a pre-concentration (or enrichment) step is required prior to any determination. Air itself is a complex mixture, being composed of gases, liquids and solid particulates; the composition of air can be influenced significantly by meteorological conditions.

Extraction Techniques in Analytical Sciences   John R. Dean
© 2009 John Wiley & Sons, Ltd

Table 11.1 Typical volatile organic compounds monitored in the atmosphere

1,1,1,2-Tetrachloroethane	1,2-Dichloropropane	Carbon tetrachloride	m,p-Xylene
1,1,1-Trichloroethane	1,3,5-Trimethylbenzene	Chlorobenzene	Naphthalene
1,1,2,2-Tetrachloroethane	1,3-Dichlorobenzene	Chloroform	n-Butylbenzene
1,1,2-Trichloroethane	1,3-Dichloropropane	cis,trans 1,3-Dichloropropene	n-Heptane
1,1-Dichloroethane	1,4-Dichlorobenzene	cis,trans-1,2-Dichloroethylene	n-Hexane
1,1-Dichloroethylene	1-Pentene	Dibromochloromethane	n-Octane
1,1-Dichloropropene	2,2-Dichloropropane	Dibromomethane	n-Pentane
1,2,3-Trichlorobenzene	2-Chlorotoluene	Dichloromethane	n-Propylbenzene
1,2,3-Trichloropropane	2-cis,trans-Pentene	Ethylbenzene	o-Xylene
1,2,3-Trimethylbenzene	4-Chlorotoluene	Hexachlorobutadiene	p-Isopropyltoluene
1,2,4-Trichlorobenzene	Benzene	i-Hexene	sec-tert-Butylbenzene
1,2-Dibromo-3-chloropropane	Bromobenzene	i-Octane	Styrene
1,2-Dibromoethane	Bromochloromethane	i-Pentane	Tetrachloroethene
1,2-Dichlorobenzene	Bromodichloromethane	Isoprene	Toluene
1,2-Dichloroethane	Bromoform	Isopropylbenzene	Trichloroethylene

> **SAQ 11.1**
>
> What meteorological conditions might affect the air composition?

11.2 Techniques Used for Air Sampling

A range of techniques are used to sample and pre-concentrate VOCs in air samples and include the following:

- whole air collection in containers;
- enrichment into solid sorbents;
- desorption techniques;
- on-line sampling.

Each approach will now be discussed in the following.

11.2.1 Whole Air Collection

This is the simplest approach for collecting air samples and uses bags or canisters. Samples are analysed either by direct injection into a GC instrument by using a gas-tight syringe or more often the air within the container needs to be pre-concentrated to allow measurement of the VOCs; this can be carried out by using, for example, a cold-trap or solid phase microextraction (SPME) device (see Chapter 4).

> **DQ 11.1**
>
> Review the technique of solid phase microextraction (SPME) to see how it might be applied in this situation.
>
> *Answer*
>
> Hint – refer to Chapter 4 when considering your response to this discussion question.

The most common containers for collecting the whole air samples are plastic bags (e.g. 'Tedlar', Teflon or 'aluminized Tedlar') and stainless-steel containers. The plastic bags are available in a range of sizes, from 500 ml to 100 l and can be re-used provided they are cleaned-out; cleaning takes place by repeatedly filling the bag with pure N_2 and evacuating with a slight negative pressure.

DQ 11.2

Why is it necessary to use pure N_2 in the cleaning process?

Answer

As the technique is being used for air sampling it is essential to maintain a 'clean' contaminant-free plastic bag.

Samples collected in plastic bags should be analysed within 24–48 h to prevent losses. Stainless-steel containers should be pre-treated to prevent internal surface reactivity by either a chrome–nickel oxide ('Summa passivation') or by chemically bonding a fused silica layer to the inner surface.

11.2.2 Enrichment into Solid Sorbents

11.2.2.1 Active Methods

In this approach a defined volume of an air sample is pumped through a solid adsorbent (or mixture of adsorbents), located within a tube, where the VOCs are retained (Figure 1.6, Chapter 1). The tube typically has dimensions of 3.5^{ii} with a $1/4^{ii}$ external diameter capable of sampling air at flow rates ranging from 10 to 200 ml/min. Stainless-steel tubes are manufactured specifically for thermal desorption (see Section 11.2.3). Typical adsorbents used for this approach are as follows:

- Porous organic polymers, such as 'Tenax', 'Chromosorb' and 'Porapak'.

- Graphitized carbon blacks, such as 'Carbotrap' and 'Carbograph'.

- Carbon molecular sieves, such as 'carbosieve' and 'carboxen'.

- Active charcoal.

SAQ 11.2

'Tenax' is one of the most commonly used adsorbents, but what is it?

It may be necessary to cryogenically cool the trap during sampling to retain the VOCs. Loss of trap efficiency can result from the presence of ozone and humidity; the former can lead to loss of VOCs, particularly unsaturated compounds, by reaction. The latter can be prevented by the inclusion of a moisture trap attached to the sampling tube. This is particularly important when using activated carbon as the adsorbent.

11.2.2.2 Passive Methods

The determination of VOCs by passive samplers relies on the diffusion of the compounds from the air to the inside of the sampling device. At that point, the VOCs are either trapped on the surface or within a trapping medium. The process can be described by using Fick's first law of diffusion which can be represented as follows:

$$m/(tA) = D(C_a - C_f)/L \tag{11.1}$$

where m = mass of substance that diffuses (µg), t = sampling interval (s), A = cross-sectional area of the diffusion path (cm^2), D = diffusion coefficient for the substance in air (cm^2 s^{-1}), C_a = concentration of substance in air (µg cm^{-3}), C_f = concentration of substance above the sorbent and L = diffusion path length (cm). If it is assumed that the adsorbent acts as a 'zero-sink' for the substance, then $C_f = 0$, and thus the equation can be simplified to:

$$m/(tC_a) = DA/L \tag{11.2}$$

The term '$m/(tC_a)$' is often called the uptake or sampling rate (R_s); in principle it is constant for a compound and a type of sampling device and so once determined can be used to determine the concentration of the substance in the air (C_a), from a measured mass of substance. Equation (11.2) is often further simplified to:

$$R_s = DA/L \tag{11.3}$$

DQ 11.3

How can you determine the diffusion coefficient, R_s?

Answer

Three approaches are possible:

(1) Use the published theoretical values of the diffusion coefficients [1].

(2) Experimentally determine the uptake rate coefficients based on the exposure of the sampler to a standard gas mixture in a chamber [2].

(3) Calculate the diffusion coefficient using the following equation [3]:

$$D = 10^{-3}\{T^{1.75}[(1/m_{air}) + (1/m)]^{1/2}\}/P(V_{air}^{1/3} + V^{1/3})2 \tag{11.4}$$

where T = absolute temperature (K), m_{air} = average molecular mass of air (28.97 g/mol), m = molecular mass of the compound (g/mol), P = gas phase pressure (atm), V_{air} = average molar volume of gases in air ($\sim$20.1 cm^3/mol) and V = molar volume of the compound (cm^3/mol).

Once the diffusion coefficient is determined, use Equation (11.3) to determine the uptake or sampling rate for the compound by measuring the cross-sectional area of the diffusion path and its diffusion path length.

A range of devices have been used as passive samplers but are largely based on either tubes or boxes (badges):

- 'Tube-type' samplers: characterized by a long, axial diffusion path and a low cross-sectional area resulting in relatively low sampling rates.

- 'Badge-type' samplers: characterized by a shorter diffusion path and a greater cross-sectional area resulting in higher uptake rates.

A range of commercial and non-commercial devices have been applied for passive air sampling. Schematic diagrams of generic passive samplers are shown in Figure 11.1. A recent review highlights the approaches to determine VOCs, PAHs and PCBs in indoor air [4].

11.2.3 Desorption Techniques

Adsorption of VOCs on solid sorbents is one of the most common approaches for air sampling. Once trapped, however, the VOCs need to be released for GC analysis. Two approaches are used: solvent desorption or thermal desorption. In the case of the former approach, solvent, e.g. DCM, is used to remove compounds from a sorbent. The approach can be effectively used for compounds that are thermally labile. As this approach uses solvent the possibility of contamination needs to be avoided; the extract may also need pre-concentration (see Chapter 1) due to the dilution effect that has taken place. This approach has been developed and the solvent desorption step refined to include the use of microwave-assisted extraction (Chapter 8) and pressurized fluid extraction (Chapter 7).

In thermal desorption, VOCs are desorbed from the solid support, within a stainless-steel tube, by heat and directly introduced into the GC injection port via a heated transfer line (Figure 11.2). The technique itself is 'solventless' (i.e. no organic solvents are used) and can be automated. It is important that the sample is heated in a manner that maximizes the recovery of the adsorbed compound without altering its chemical composition. In order to maintain compound integrity, relatively cool temperatures (e.g. 100°C) are used; unfortunately the desorption of compounds at these temperatures may be slow. This results in the compounds having broad, poorly resolved peaks in the chromatogram. However, this is not always the case, and some compounds will desorb rapidly, producing good peak shape. An approach to prevent poor GC resolution is to trap the VOCs cryogenically onto the GC column before initializing the temperature programme. This can be achieved by utilizing the GC oven's cryogenic function

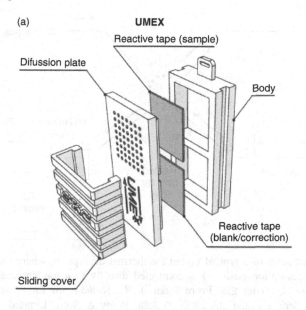

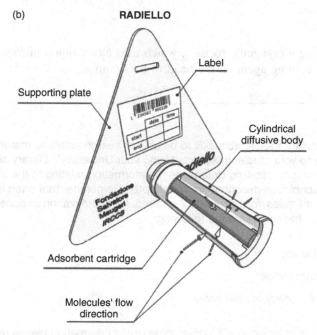

Figure 11.1 Passive sampling using (a) a 'tube-type' sampler and (b) a 'badge-type' sampler. Reprinted from *Anal. Chim. Acta*, **602**(2), Kot-Wasik *et al.*, 'Advances in passive sampling in environmental studies', 141–163, Copyright (2007) with permission from Elsevier.

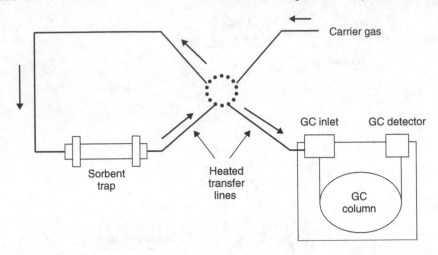

Figure 11.2 Illustration of a typical layout for thermal desorption, where the desorption unit (set in the desorption position) is connected directly to a gas chromatograph: → indicates the flow of carrier gas. From Dean, J. R., *Methods for Environmental Trace Analysis*, AnTS Series. Copyright 2003. © John Wiley & Sons, Limited. Reproduced with permission.

or by installing a cryogenic focuser, which uses either liquid nitrogen or carbon dioxide as a cooling agent, at the head of the column.

SAQ 11.3

It is an important transferable skill to be able to search scientific material of importance to your studies/research. Using your University's Library search engine search the following databases for information relating to the air sampling techniques described in this chapter. Remember that often these databases are 'password-protected' and require authorization to access. Possible databases include the following:

- Science Direct;
- Web of Knowledge;
- The Royal Society of Chemistry.

(While the use of 'google' will locate some useful information please use the above databases.)

Summary

A whole range of approaches for recovering organic compounds from air samples is available. This chapter describes each of these approaches, highlighting the key principles and aspects of the techniques. A review of the air sampling approaches highlights the diversity of the applications.

References

1. Lide, D. R. (Ed.), *CRC Handbook of Chemistry and Physics*, 86th Edn, CRC Press, Boca Raton, FL, USA (2005).
2. Partyka, M., Zabiegala, B., Namiesnik, J. and Przyjazny, A., *Crit. Rev. Anal. Chem.*, **37**, 51–78 (2007).
3. Schwarzenbach, R.P., Gschwend, P. M. and Imboden, D. M., *Environmental Organic Chemistry*, Wiley-VCH, New York, NY, USA (1993).
4. Barro, R., Regueiro, J., Llompart, M. and Garcia-Jares, C., *J. Chromatogr., A*, **1216**, 540–566 (2009).

COMPARISON OF EXTRACTION METHODS

Chapter 12

Comparison of Extraction Methods

Learning Objectives

- This chapter outlines the main considerations in the selection of an extraction technique for recovering organic compounds from solid, aqueous and air samples.
- The role of certified reference materials in the laboratory aspects of extraction/analysis is highlighted.
- Suppliers of these materials are also provided.

12.1 Introduction

Any comparison of different extraction methods is difficult to determine as it is requires the selection of key parameters of importance to the user. Obviously these may vary between different users.

DQ 12.1

Suggest appropriate extraction method criteria that allow a direct comparison.

Answer

The following may be appropriate criteria:

Extraction Techniques in Analytical Sciences   John R. Dean
© 2009 John Wiley & Sons, Ltd

- *Sample mass/volume.* The amount of sample that an extraction technique requires is an important aspect and can directly influence the sensitivity of the measurement component – more analyte that can be extracted from a larger sample will allow the measurement of the analyte to be made at a lower concentration.

- *Extraction time.* The length of time that the extraction methodology takes is one important consideration. However, while it may be obvious to link the extraction time (faster is better) with the analysis step the argument does not always hold. Just as multiple samples can be extracted simultaneously, using some approaches, so the use of 'auto samplers' on chromatographic systems means that multiple sample extracts can be pre-loaded ready for analysis overnight, if necessary. Perhaps the faster extraction is better assessed in terms of the customer requirements/needs.

- *Solvent type and consumption.* Not all extraction techniques require solvent as part of the process. If solvent is required it would be beneficial if the type of solvent used could be environmentally friendly, cheap to purchase with minimal disposal cost and that small quantities could be used.

- *Extraction method.* A range of approaches exist for the recovery of analytes from (semi)-solid, liquid and air samples. The dilemma is to assess which approach best suits your needs/requirements. This may not be easy as most research scientists rely on commercial extraction techniques, often available from a range of suppliers.

- *Sequential or simultaneous extraction.* This criterion could be taken alongside the 'extraction time' criterion above. However, the question is more fundamental. Is it better to extract a sample using a 'one-at-a-time' approach or to extract samples 'several-at-a-time'? The latter is undoubtedly important once any experimental variation in the influence of the extraction technique is known and can be simply repeated multi-fold. The sequential approach does provide some investigation of the important operating variables of the extraction technique/methodology. An understanding of these variables could have long-term benefits, if properly understood.

- *Method development time.* Ideally this should be as short as possible. For research scientists in academia this could lead to a journal publication but in the commercial sector this is costly and perhaps unproductive.

- *Operator skill.* No one would want an extraction technique that requires a high level of operator skill to operate, at least not on a routine basis. Highly skilled operators may be required to assess variable/parameter influence on extraction recovery. However, once the approach has been developed the process should be capable of being operated routinely. The more complicated a system is to use, the more likely it is to lead to worse precision. Maintenance of the extraction technique is also an important consideration. The more complicated the extraction technique, the more highly skilled the operative is required to be to ensure its safe and continued operation.

- *Equipment cost.* No one wants to pay a large amount of money for the extraction approach adopted provided the chosen one is effective and in-line with other customer/client criteria. Nevertheless all approaches have an inherent capital cost that needs to be assessed as part of their selection criteria. In addition to the initial capital outlay cost it is also important to consider the routine and regular cost for maintaining the extraction technique in consumables and maintenance costs.

- *Level of automation.* The greater the level of automation, the undoubted higher the initial capital cost and possibly the higher routine running costs. However, these costs may be overcome by (a) the lower costs in terms of staffing that may be required or (b) the deployment of staff on more productive aspects rather than routine activities.

- *Extraction method approval.* Several organizations worldwide produce 'methods' that have been tested and 'approved' for use in extraction analytes from matrices. The most comprehensive list of 'official' environmental methods has been produced by the US Environmental Protection Agency (USEPA). Other organizations that produce 'approved' methods include the following: Association of Official Analytical Chemists (AOAC); Deutsches Institut für Normung (DIN); National Metrology Institute of Japan (NMIJ); American Society for Testing and Materials (ASTM).

12.2 Role of Certified Reference Materials

The use of any extraction technique requires some verification that the approach is effective, reliable, reproducible and accurate. Obviously the use of an extraction technique is only part of the process and it is therefore impossible to ignore the analysis stage in any protocol evaluation. Nevertheless the use of Certified Reference Materials (CRMs) provides an opportunity to assess the overall extraction–analysis process in terms of its reliability. In selecting a matrix reference material (i.e. one in which a specific analyte or range of analytes is located within a named and specific matrix) it is important to consider the following:

- *Matrix match.* It is important to select a CRM with a similar matrix to the sample itself. The choice of a soil CRM may not be so specific, particularly if the extraction technique has some dependency upon soil organic matter content. It may be necessary to select a 'sandy, loam soil' CRM, for instance, to be compatible with the soil under investigation.

- *Analytes.* It is common to be extracting and then analysing a range of related analytes in the sample, e.g. polycyclic aromatic hydrocarbons (PAHs), pesticides etc. On that basis it is necessary to include in the CRM selection process the most appropriate reference material/analyte combination.

- *Measurement range.* As well as selecting the range of analytes in a specific sample matrix for the CRM it is also necessary to consider the measurement range of the analyte(s). In order to have confidence after the extraction/analysis of reliable sample data it is necessary to be using a CRM with a similar measurement range. For example, it is unreliable to be using a CRM with certified values in the mg/kg range for your specific analytes when you are extracting/analysing in the µg/kg range.

- *Measurement uncertainties.* As the purpose of the CRM is to allow the user to achieve the measurement concentration within a given uncertainty it is necessary to give some thought to the expected measured uncertainties. If the quoted measured uncertainties are so large that poor laboratory practice will allow values to be obtained within their limits then the use of such material needs to be questioned. The best CRM values should have measurement ranges and uncertainties that are achievable by the majority of users provided they are operating good laboratory practice protocols and that the procedures adopted for extraction/analysis are appropriately carried out.

- *Certification procedures used by the CRM producer.* The producer will indicate how the sample was extracted/analysed (which may be the same as you, the user of the CRM).

- *Documentation supplied with the material.* Every sample purchased will arrive with documentation indicating the following (as a minimum): information on how the sample was prepared, minimum sample size, whether dry weight is important (and hence necessary to consider in the analytical protocol) and shelf-life. This documentation will list the analytes present in the sample, together with either a given uncertainty (if certified) or an indicative value, per analyte.

The most common suppliers of CRMs are:

- The National Institute of Standards and Technology (NIST), USA [www.NIST.org].

- Laboratory of the Government Chemist (LGC), UK [http://www.lgc.co.uk].

- Institute for Reference Materials and Measurements (IRMM), Belgium [www.IRMM.org].

Table 12.1 Types of compounds in certified reference materials

Compound abbreviation	Name of compound(s)
PCBs	Polychlorinated biphenyls
PAHs	Polycyclic aromatic hydrocarbons
PCP	Pentachlorophenol
PCDDs	Polychlorinated dibenzo-*p*-dioxins
PCDFs	Polychlorinated dibenzofurans
BTEX	Benzene, toluene, ethylbenzene and xylenes
TPHs	Total petroleum hydrocarbons
VOCs	Volatile organic compounds
VOAs	Volatile organic analytes
BNAs	Base, neutral and acidic compounds

- The Federal Institute for Materials Research and Testing (BAM), Germany [www.bam.de].

- National Metrology Institute of Japan (NMIJ) [http://www.nmij.jp].

- The National Research Council of Canada (NRC) [http://www.nrc-cnrc.gc.ca].

- The National Water Research Institute (NWRI), Canada [http://www.ec.gc.ca].

- National Research Centre for Certified Reference Materials (NRCCRM), China [http://www.nrccrm.org.cn].

- RT Corporation, USA [http://www.rt-corp.com].

The main groups of compounds for which CRMs have been produced are shown in Table 12.1.

12.3 Comparison of Extraction Techniques for (Semi)-Solid Samples

It is possible to compare the advantages and disadvantages of Soxhlet, shake-flask, sonication, matrix solid phase dispersion (MSPD), SFE and MAE with PFE using the above criteria (see Section 12.1). Such a comparison is shown in Table 12.2.

SAQ 12.1

Using the criteria identified in DQ 12.1 (above) compare the criteria for extraction of organic compounds from solid matrices.

Table 12.2 Comparison of extraction techniques for recovery of organic compounds from solid sample matrices[a]

Feature	Soxhlet	Shake-flask	Sonication	MSPD	SFE	MAE	PFE
Sample mass	Normally up to 10 g	0.5–10 g	2–30 g	0.5–10 g	1–10 g	2–10 g	Normally up to 30 g
Extraction time	6, 12 or 24 h per sample	3–5 min per sample	3–5 min per sample	5–20 min per sample	30 min to 1 h per sample	20 min (plus 30 min cooling and pressure reduction) for up to 40 samples	12–15 min per sample
Solvent type	Acetone/hexane (1:1, vol/vol); acetone/DCM (1:1, vol/vol); DCM only; toluene/methanol (10:1, vol/vol)	Typically, acetone/DCM (1:1, vol/vol)	Acetone/DCM (1:1, vol/vol) or acetone/hexane (1:1, vol/vol) for semi-volatile organics and OCPs; acetone/DCM (1:1, vol/vol), acetone/hexane (1:1, vol/vol) or hexane for PCBs	Typically, DCM, hexane, ethyl acetate, acetonitrile, methanol or acetone	CO_2 (plus organic modifier). Tetra-chloroethene used as the collection solvent for TPHs for determination by FTIR, otherwise DCM	Typically, acetone/hexane (1:1, vol/vol). The solvent(s) is/are required to be able to absorb microwave energy	Acetone/hexane (1:1, vol/vol) or acetone/DCM (1:1, vol/vol) for OCPs, semi-volatile organics, PCBs or OPPs; acetone/DCM/phosphoric acid (250:125:15, vol/vol/vol) for chlorinated herbicides
Solvent consumption	150–300 ml	5–20 ml	5–20 ml	5–50 ml	10–20 ml	25–45 ml	25 ml
Extraction method	Heat	Agitation	Ultrasound	Solid phase extraction	Heat + pressure	Heat + pressure	Heat + pressure

	Sequential (but multiple assemblies can operate simultaneously)	Sequential (but possible to shake several flasks at the same time)	Sequential (but possible to sonicate several flasks at the same time)	Sequential (but possible to use a manifold vacuum system for up to 12 samples at the same time)	Sequential	Simultaneous (up to 40 vessels can be extracted simultaneously)	Sequential + simultaneous (up to 6 vessels can be extracted simultaneously)
Sequential or simultaneous							
Method development time	Low	Low	Low	Medium	High	High	High
Operator skill	Low	Low	Low	Medium	High	Moderate	Moderate
Equipment cost	Low	Low	Low	Low	High	Moderate	High
Level of automation	Minimal	Minimal	Minimal	Minimal	Minimal to high	Minimal	Can be fully automated
USEPA method	3540	—	3550	—	3560 for TPHs, 3561 for PAHs and 3562 for PCBs and OCPs	3546	3545

*a*MSPD, matrix solid phase dispersion; SFE, supercritical fluid extraction; MAE, microwave-assisted extraction; TPHs, total petroleum hydrocarbons; PAHs, polycyclic aromatic hydrocarbons; OCPs, organochlorine pesticides; DCM, dichloromethane (or methylene chloride); USEPA, United States of America Environmental Protection Agency.

12.3.1 A Comparison of Extraction Techniques for Solid Samples: a Case Study [1]

As part of a certification process for two sediment CRMs, a thorough investigation, by the National Metrology Institute of Japan, into a range of extraction techniques has been published [1].

The organic compounds to be determined were a range of PCBs and OCPs in two sediments (NMIJ CRM 7304a and 7305a). Specifically, the PCB congeners (PCB numbers 3, 15, 28, 31, 70, 101, 105, 138, 153, 170, 180, 194, 206 and 209), plus the OCPs (γ-HCH, 4,4'-DDT, 4,4'-DDE and 4,4'-DDD). The levels of pollutants in NMIJ CRM 7304a are higher (between 2 and 15 times greater) than in NMIJ CRM 7305a. The extraction techniques used were all multiple extraction techniques: PFE, MAE, saponification, Soxhlet, SFE and ultrasonic extraction. Following extraction, sample extracts were cleaned-up prior to determination by isotope dilution–gas chromatography–mass spectrometry (ID–GC–MS). The analytical protocol schemes for the extraction of PCBs and OCPs from the two sediment CRMs are shown in Figures 12.1 and 12.2, respectively. Each figure indicates the following:

- Extraction technique to be used (saponification will not be discussed as it has not been discussed previously in this book).

- Choice of solvent or solvents used for the specific extraction technique.

- Clean-up procedures adopted.

- Specific fractions isolated, as appropriate.

- Column used for analytical separation.

- Analytical technique used, i.e. ID–GC–MS.

- Method number for identification purposes.

It is worth noting the extensive clean-up procedures adopted for Soxhlet, PFE, MAE and ultrasonic extractions when compared to SFE.

Optimal extraction conditions were determined for the recovery of PCBs and OCPs from sediments and these are shown in Table 12.3.

The results for PCBs and OCPs in NMIJ CRM 7304a are shown in Tables 12.4 and 12.5, respectively, whereas for PCBs and OCPs in NMIJ CRM 7305a the results are shown in Tables 12.6 and 12.7, respectively. It can be seen that the data obtained are comparable, irrespective of the extraction technique used, the organic compounds investigated and the requirements for clean-up (or not). Finally, the NMIJ published their data indicating the levels of PCBs and OCPs in the two CRMs (Table 12.8).

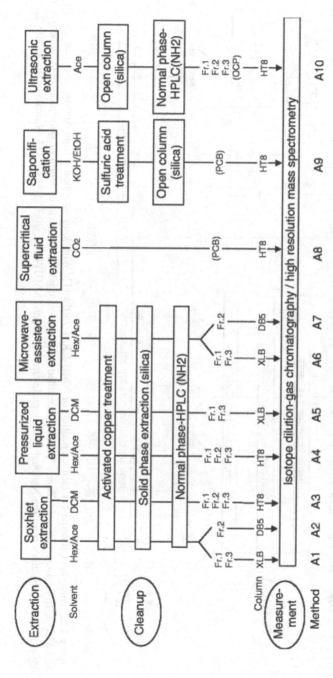

Figure 12.1 Analytical scheme for characterization of NMIJ CRM 7304a [1]. With kind permission from Springer Science and Business Media, from *Anal. Bioanal. Chem.*, 'Sediment certified reference materials for the determination of polychlorinated biphenyls and organochlorine pesticides from the National Metrology Institute of Japan (NMIJ)', **387**, 2007, 2313–2323, Numata *et al.*, Figure 1.

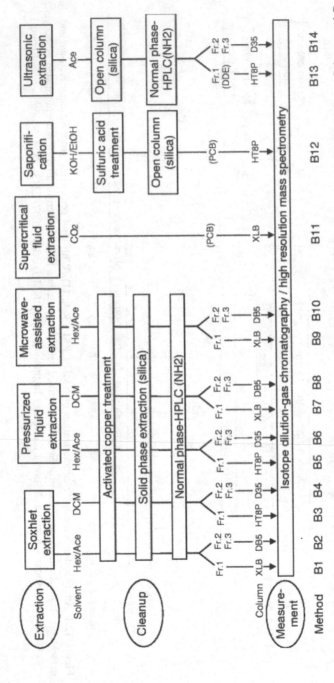

Figure 12.2 Analytical scheme for characterization of NMIJ CRM 7305a [1]. With kind permission from Springer Science and Business Media, from *Anal. Bioanal. Chem.*, 'Sediment certified reference materials for the determination of polychlorinated biphenyls and organochlorine pesticides from the National Metrology Institute of Japan (NMIJ)', **387**, 2007, 2313–2323, Numata *et al.*, Figure 2.

Table 12.3 Optimal extraction conditions for the techniques investigated [1]. With kind permission from Springer Science and Business Media, from *Anal. Bioanal. Chem.*, 'Sediment certified reference materials for the determination of polychlorinated biphenyls and organochlorine pesticides from the National Metrology Institute of Japan (NMIJ)', **387**, 2007, 2313–2323, Numata *et al.*, Table 1[a]

Technique	Solvent	Conditions
Soxhlet extraction	Hex/Ace (1:1) or DCM	Reflux, 24 h
Pressurized liquid extraction	Hex/Ace (1:1) or DCM	150°C, 15 MPa, 30 min × 2 cycles
Microwave-assisted extraction	Hex/Ace (1:1)	145°C, 20 min
Supercritical fluid extraction	CO_2 (no modifier)	140°C, 30 MPa, 15 min (static) → 30 min (dynamic)
Saponification	1 M KOH/EtOH → Hex	Room temp., shake 1 h → (residue) → 80°C, reflux, 1 h

[a] Hex, hexane; Ace, acetone; DCM, dichloromethane; EtOH, ethanol.

12.4 Comparison of Extraction Techniques for Liquid Samples

It is possible to compare the advantages and disadvantages of Soxhlet, shake-flask, sonication, matrix solid phase dispersion (MSPD), SFE and MAE with PFE using the above criteria (see Section 12.1). The comparison is shown in Table 12.9.

12.5 Comparison of Extraction Techniques for Air Sampling

A range of approaches are available for air sampling and range from whole air sampling using Tedlar bags or stainless-steel canisters through to compound enrichment/pre-concentration on sorbents via either active or passive sampling.

SAQ 12.2

It is an important transferable skill to be able to search scientific material of importance to your studies/research. Using your University's Library search engine search the following databases for information relating to scientific papers or reviews that compare extraction techniques. Remember that often these databases are 'password-protected' and require authorization to access. Possible databases include the following:

(*continued on p. 238*)

Table 12.4 Analytical results for the determination of PCB congeners in CRM 7304-a [1]. With kind permission from Springer Science and Business Media, from *Anal. Bioanal. Chem.*, 'Sediment certified reference materials for the determination of polychlorinated biphenyls and organochlorine pesticides from the National Metrology Institute of Japan (NMIJ)', **387**, 2007, 2313–2323, Numata *et al.*, Table 2[a]

	Method A1	Method A3	Method A4	Method A5	Method A6	Method A8	Method A9
PCB3	0.285 (0.0123)	0.281 (0.0098)	0.351 (0.0097)	0.392 (0.0189)	0.305 (0.0071)	0.348 (0.0226)	0.280 (0.0077)
PCB15	2.23 (0.085)	2.21 (0.072)	2.36 (0.070)	2.37 (0.052)	2.33 (0.044)	2.16 (0.075)	2.08 (0.064)
PCB28	34.8 (0.67)	33.3 (0.90)	35.5 (0.84)	34.9 (0.59)	35.0 (0.62)	34.0 (0.96)	36.1 (0.80)
PCB31	27.4 (1.04)	26.0 (0.64)	27.5 (0.68)	27.6 (1.02)	27.3 (1.02)	26.6 (0.71)	28.0 (0.62)
PCB70	60.7 (1.42)	60.1 (1.21)	62.0 (1.56)	61.1 (1.50)	60.5 (1.41)	58.2 (1.52)	62.2 (1.20)
PCB101	32.5 (0.71)	30.9 (0.82)	32.8 (1.05)	32.7 (0.66)	32.2 (0.68)	29.8 (0.89)	32.8 (0.75)
PCB105	18.4 (0.44)	17.5 (0.52)	18.8 (0.43)	18.9 (0.50)	18.2 (0.42)	18.2 (0.59)	18.9 (0.49)
PCB138	14.1 (0.37)	13.4 (0.32)	14.1 (0.47)	14.4 (0.37)	14.1 (0.36)	13.1 (0.40)	13.9 (0.32)
PCB153	16.2 (0.52)	15.7 (0.42)	16.3 (0.47)	16.5 (0.55)	16.0 (0.53)	15.2 (0.42)	16.0 (0.36)
PCB170	3.53 (0.13)	3.51 (0.14)	3.71 (0.13)	3.70 (0.11)	3.56 (0.11)	3.64 (0.15)	3.74 (0.12)
PCB180	8.93 (0.32)	8.88 (0.39)	9.52 (0.28)	9.15 (0.26)	8.67 (0.23)	9.07 (0.35)	9.55 (0.31)
PCB194	1.81 (0.079)	1.89 (0.110)	1.99 (0.095)	1.91 (0.069)	1.84 (0.059)	1.90 (0.120)	1.96 (0.105)
PCB206	0.467 (0.044)	0.454 (0.040)	0.476 (0.041)	0.507 (0.049)	0.475 (0.042)	0.472 (0.044)	0.486 (0.045)
PCB209	1.32 (0.153)	1.16 (0.129)	1.26 (0.078)	1.56 (0.133)	1.35 (0.133)	1.25 (0.112)	1.19 (0.058)

[a]The unit of values is µg kg^{-1} dry mass. Values in parentheses are $u(C$ ind) (i.e. uncertainties associated with each analytical method).

Table 12.5 Analytical results for the determination of OCPs in CRM 7304-a [1]. With kind permission from Springer Science and Business Media, from *Anal. Bioanal. Chem.*, 'Sediment certified refernece materials for the determination of polychlorinated biphenyls and organochlorine pesticides from the National Metrology Institute of Japan (NMIJ)', **387**, 2007, 2313–2323, Numata et al., Table 3[a]

	Method A1	Method A2	Method A3	Method A4	Method A5	Method A6	Method A7	Method A10
4,4′-DDT	–	5.89 (0.278)	5.28 (0.253)	–	–	–	5.30 (0.263)	5.36 (0.189)
4,4′-DDE	5.37 (0.186)	–	5.42 (0.187)	5.57 (0.508)	5.40 (0.438)	5.39 (0.103)	–	5.24 (0.135)
4,4′-DDD	12.7 (0.44)	–	11.3 (0.21)	13.3 (0.30)	14.0 (0.45)	12.5 (0.21)	–	11.8 (0.18)
γ-HCH	5.63 (0.342)	–	5.15 (0.190)	5.57 (0.394)	5.38 (0.172)	5.22 (0.130)	–	3.92 (0.091)[b]

[a]The unit of values is μg kg^{-1} dry mass. Values in parentheses are $u(C$ ind) (i.e. uncertainties associated with each analytical method).
[b]Concentration of γ-HCH obtained by method A10 was not used for estimation of the certified value but for the information value.

Table 12.6 Analytical results for the determination of PCB congeners in CRM 7305-a [1]. With kind permission from Springer Science and Business Media, from *Anal. Bioanal. Chem.*, 'Sediment certified refernece materials for the determination of polychlorinated biphenyls and organochlorine pesticides from the National Metrology Institute of Japan (NMIJ)', **387**, 2007, 2313–2323, Numata *et al.*, Table 4[a]

	Method B1	Method B3	Method B5	Method B7	Method B9	Method B11	Method B12
PCB3	0.1104 (0.0057)	0.1181 (0.0102)	0.1838 (0.0107)	0.1874 (0.0054)	0.1662 (0.0061)	–	0.1242 (0.0096)
PCB15	0.2952 (0.0120)	0.2881 (0.0128)	0.3320 (0.0068)	0.3490 (0.0107)	0.3354 (0.0104)	0.2838 (0.0208)	0.2859 (0.0067)
PCB28	2.852 (0.078)	2.860 (0.071)	2.975 (0.068)	2.923 (0.078)	2.951 (0.081)	2.766 (0.084)	2.736 (0.073)
PCB31	2.263 (0.070)	2.214 (0.052)	2.318 (0.053)	2.353 (0.074)	2.370 (0.082)	2.200 (0.075)	2.149 (0.052)
PCB70	3.960 (0.130)	4.060 (0.060)	4.117 (0.066)	4.024 (0.085)	4.070 (0.089)	3.813 (0.087)	3.968 (0.077)
PCB101	2.805 (0.136)	2.439 (0.053)	2.561 (0.049)	2.628 (0.085)	2.727 (0.078)	2.474 (0.090)	2.488 (0.067)
PCB105	1.313 (0.067)	1.270 (0.037)	1.257 (0.034)	1.254 (0.024)	1.307 (0.023)	1.237 (0.030)	1.233 (0.031)
PCB138	2.076 (0.119)	1.844 (0.029)	1.888 (0.031)	2.004 (0.110)	1.999 (0.080)	1.906 (0.094)	2.023 (0.105)
PCB153	3.360 (0.134)	3.037 (0.051)	3.113 (0.038)	3.396 (0.140)	3.273 (0.058)	3.129 (0.072)	3.442 (0.218)
PCB170	0.9790 (0.052)	0.8273 (0.056)	0.8472 (0.024)	1.045 (0.065)	0.9338 (0.034)	0.9026 (0.029)	1.025 (0.093)
PCB180	2.559 (0.202)	2.290 (0.075)	2.299 (0.041)	2.805 (0.167)	2.493 (0.094)	2.360 (0.065)	2.865 (0.280)
PCB194	0.6600 (0.045)	0.5663 (0.017)	0.5663 (0.014)	0.7160 (0.038)	0.6364 (0.037)	0.6294 (0.021)	0.7154 (0.059)
PCB206	0.1493 (0.016)	0.1324 (0.011)	0.1450 (0.009)	0.1559 (0.014)	0.1541 (0.015)	0.1498 (0.012)	0.1604 (0.015)
PCB209	0.1625 (0.014)	0.1701 (0.014)	0.1815 (0.014)	0.1814 (0.014)	0.1485 (0.010)	0.1561 (0.014)	0.1463 (0.009)

[a]The unit of values is µg kg^{-1} dry mass. Values in parentheses are $u(C$ ind) (i.e. uncertainties associated with each analytical method).

Table 12.7 Analytical results for the determination of OCP congeners in CRM 7305-a [1]. With kind permission from Springer Science and Business Media, from *Anal. Bioanal. Chem.*, 'Sediment certified refernece materials for the determination of polychlorinated biphenyls and organochlorine pesticides from the National Metrology Institute of Japan (NMIJ)', **387**, 2007, 2313–2323, Numata *et al.*, Table 5[a]

	Method B2	Method B4	Method B6	Method B8	Method B10	Method B14
4,4'-DDT	2.178 (0.088)	1.750 (0.105)	2.134 (0.116)	2.527 (0.223)	2.414 (0.157)	2.375 (0.203)
4,4'-DDD	3.259 (0.100)	3.118 (0.111)	3.408 (0.115)	3.454 (0.128)	3.520 (0.083)	3.176 (0.121)
γ-HCH	0.8512 (0.036)	0.9353 (0.046)	0.8772 (0.033)	1.008 (0.041)	0.8252 (0.031)	0.5522 (0.025)
	Method B1	Method B3	Method B5	Method B7	Method B9	Method B13
4,4'-DDE	1.748 (0.040)	1.762 (0.032)	1.831 (0.027)	1.794 (0.046)	1.888 (0.058)	1.772 (0.026)[b]

[a]The unit of values is μg kg^{-1} dry mass. Values in parentheses are $u(C$ ind) (i.e. uncertainties associated with each analytical method).
[b]Concentration of γ-HCH obtained by method A10 was not used for estimation of the certified value but for the information value.

Table 12.8 Certified values for organic pollutants in NMIJ CRM 7304-a and CRM 7305-a [1]. With kind permission from Springer Science and Business Media, from *Anal. Bioanal. Chem.*, 'Sediment certified reference materials for the determination of polychlorinated biphenyls and organochlorine pesticides from the National Metrology Institute of Japan (NMIJ)', **387**, 2007, 2313–2323, Numata *et al.*, Table 6[a]

	Certified value (mass fraction, $\mu g\ kg^{-1}$ dry mass)	
	NMIJ CRM 7304-a	NMIJ CRM 7305-a
PCB congeners		
PCB3	0.311 ± 0.085	0.15 ± 0.07
PCB15	2.26 ± 0.24	0.31 ± 0.05
PCB28	34.9 ± 2.3	2.9 ± 0.2
PCB31	27.1 ± 1.8	2.26 ± 0.18
PCB70	60.7 ± 3.8	4.0 ± 0.3
PCB101	31.9 ± 2.6	2.6 ± 0.3
PCB105	18.4 ± 2.0	1.27 ± 0.14
PCB138	13.9 ± 1.1	1.92 ± 0.15
PCB153	15.9 ± 1.0	3.2 ± 0.3
PCB170	3.62 ± 0.22	0.92 ± 0.16
PCB180	9.10 ± 0.69	2.4 ± 0.5
PCB194	1.89 ± 0.11	0.62 ± 0.13
PCB206	0.476 ± 0.050	0.15 ± 0.03
PCB209	1.28 ± 0.20	0.16 ± 0.03
Organochlorine pesticides		
4,4'-DDT	5.44 ± 0.50	2.2 ± 0.5
4,4'-DDE	5.37 ± 0.30	1.79 ± 0.11
4,4'-DDD	12.4 ± 1.9	3.3 ± 0.3
γ-HCH	5.33 ± 0.26	0.89 ± 0.12

[a] Results are expressed as the certified concentration $\pm$ expanded uncertainty ($k = 2$).

(*continued from p. 233*)

- Science Direct;

- Web of Knowledge;

- The Royal Society of Chemistry.

(While the use of 'google' will locate some useful information please use the above databases.)

Summary

This chapter outlines the main considerations in the selection of an extraction technique for recovering organic compounds from solid, aqueous and air samples.

Table 12.9 Comparison of extraction techniques for recovery of organic compounds from liquid sample matrices

Feature	Liquid–liquid	Purge and trap	SPE	SPME	Sorption method
Sample volume	Up to 1 l	5–25 ml	1 ml to 1 l	1 ml to 1 l	1 ml to 1 l
Extraction time	Discontinuous, 20 min; continuous, up to 24 h	10–20 min	10–20 min	2–60 min	2–60 min
Solvent type	Various organic solvents	Nitrogen for purging/desorption	Various (depending upon nature of sorbent phase)	No solvent	No solvent/minimal
Solvent consumption	3 × 60 ml for discontinuous; up to 500 ml for continuous	No solvent	Organic solvent required for wetting sorbent, clean-up stage and elution step (up to 20–30 ml)	None	None/minimal
Extraction method	Partitioning into organic phase	Desorption into gas phase	Partitioning onto sorbent	Partitioning onto sorbent	Partitioning onto sorbent
Sequential or simultaneous	Sequential	Sequential	Sequential and simultaneous	Sequential	Sequential
Method development time	Low	Low	Low–moderate	Low–moderate	Moderate
Operator skill	Low	Moderate	Low–moderate	Low–moderate	Moderate
Equipment cost	Low	Moderate	Low–high	Low	Low
Level of automation	Low	Moderate	Low–high (robotic systems)	Low–moderate (autosamplers)	Low–moderate
USEPA method	Methods 3510 and 3520	Method 5030	Method 3535	None	None

The role of Certified Reference Materials in the laboratory aspects of extraction/analysis is highlighted. Suppliers of these materials are also highlighted.

References

1. Numata, M., Yarita, T., Aoyagi, Y., Tsuda, Y., Yamazaki, M., Takatsu, A., Ishikawa, K., Chiba, K. and Okamaoto, K., *Anal. Bioanal. Chem*., **387**, 2313–2323 (2007).

RESOURCES

Chapter 13

Resources for Extraction Techniques

Learning Objectives

- To be able to identify appropriate resources to maintain an effective knowledge of development in this subject matter.

13.1 Introduction

It is important to keep-up-to-date in the area of extraction techniques in analytical sciences to ensure that the latest developments in techniques and applications are known, so as to influence your research and/or study being undertaken. However, it is virtually impossible to be able to consider everything in 'hard' and 'electronic' copies (unless that is your sole occupation!). So how can you tackle the vast amount of information that is available?

Here are some general tips to consider:

- Accept that you cannot access all information and develop your strategy to assimilate relevant key data.

- What are the sources of the relevant data?

- How will you seek to obtain this information?

Extraction Techniques in Analytical Sciences   John R. Dean
© 2009 John Wiley & Sons, Ltd

- How will you assess whether the content of the sourced information is relevant?

- How will you seek to modify the information and apply it in your work?

13.1.1 Sources of Data

The most common sources of information for an individual is via journals, books, conferences and manufacturers/suppliers. However, the quantity of material produced in terms of this subject matter is enormous and needs to be targeted. For example, no one is going to read all relevant journals! So the first key objective is to identify the most relevant journals which publish material that is of interest and relevance to you and your work/research. Some journals in this field are generic and publish widely in analytical chemistry, e.g. *Analytical Chemistry, The Analyst* and *Analytica Chimica Acta*, while other journals focus on techniques, e.g. *Journal of Chromatography, A* and *B*, with others on specific applications, e.g. *Environmental Science and Technology* and *Environmental Pollution*. Once you have identified your key journals it is then possible to obtain the journal contents for free by signing up for their respective 'e-mail altering services', thus allowing the latest publications in a particular field of study to be directly forwarded to you (via e-mail). Some selected web sites for the major publishers are given in Table 13.1.

Most journals are also available electronically on your desktop PC subject to the necessary payment being made. Payment of the subscription fee is often carried out by libraries in universities, industry or public organizations. Electronic access to journals allows the full text to be read in either PDF or HTML formats. In the former case, i.e. PDF format, the article appears in exactly the same format as the print copy, while in the latter case, i.e. HTML format, the article will have weblinks (i.e. hyperlinks) to tables, figures or references (the references themselves are often further linked to their original sources by using a 'reference-linking' service).

13.2 Role of Worldwide Web

To gain access to the Internet requires the use of a web browser, e.g. 'Microsoft Internet Explorer'. Searching the web for useful information is carried out via a search engine, e.g. 'Google'. It should be remembered that searching the web can be very time-consuming. Therefore browsing should be focused on relevant and specific sites.

Some of the main resources you can utilize via the web are as follows:

- *Publishers.* These provide access to their catalogues of journals (Table 13.1) and books 'on-line' (e.g. Wiley (www.wiley.com) and Pearson (www. pearsonhighered.com)). Access to browse and search the databases of articles

Table 13.1 Alphabetical list of selected journals that publish articles (research papers, communications, critical reviews, etc.) on analytical techniques and their applications

Journal	Publisher	Web address[a]
Analyst	The Royal Society of Chemistry	http://www.rsc.org/Publishing/Journals/an/
Analytica Chimica Acta	Elsevier	http://www.elsevier.com/wps/find/journaldescription.cws_home/502681/description#description
Analytical Chemistry	American Chemical Society	http://pubs.acs.org/journals/ancham/index.html
Chemosphere	Elsevier	http://www.elsevier.com/wps/find/journaldescription.cws_home/362/description#description
Environmental Pollution	Elsevier	http://www.elsevier.com/wps/find/journaldescription.cws_home/405856/description#description
Environmental Science and Technology	American Chemical Society	http://pubs.acs.org/journals/esthag/
Journal of Agricultural and Food Chemistry	American Chemical Society	http://pubs.acs.org/journals/jafcau/index.html
Journal of Chromatography, A	Elsevier	http://www.elsevier.com/wps/find/journaldescription.cws_home/502688/description#description
Journal of Environmental Monitoring (JEM)	The Royal Society of Chemistry	http://www.rsc.org/Publishing/Journals/em/index.asp
Microchemical Journal	Elsevier	http://www.elsevier.com/wps/find/journaldescription.cws_home/620391/description#description
Science of the Total Environment	Elsevier	http://www.elsevier.com/wps/find/journaldescription.cws_home/503360/description#description
Talanta	Elsevier	http://www.elsevier.com/wps/find/journaldescription.cws_home/525438/description#description
Trends in Analytical Chemistry	Elsevier	http://www.elsevier.com/wps/find/journaldescription.cws_home/502695/description#description

[a]As of April 2009. The products or material displayed are not endorsed by the author or the publisher of this present text.

Table 13.2 Selected suppliers of instrumental extraction apparatus[a]

Suppliers of PFE Equipment
Applied Separations (www.appliedseparations.com)
Dionex Corporation (www.dionex.com)
Fluid Management Systems (www.fmsenvironmental.com)

Suppliers of MAE Equipment
Anton–Parr (www.anton-paar.com)
CEM Corporation (www.cem.com)
Milestone (www.milestonesci.com)

Suppliers of SFE Equipment
Applied Separations (www.appliedseparations.com)
Separex (www.separex.fr/) – process SFE systems
TharSFC (www.tharsfc.com/) – supercritical fluid chromatography systems

Selected other Suppliers of Extraction Equipment and Consumables
Agilent (www.home.agilent.com/)
Gerstel (www.gerstel.com/) for stir-bar sorptive extraction (SBSE)
Millipore (www.millipore.com/)
Phenonemex (www.phenomenex.com)
SGE (www.sge.com) for microextraction in a packed syringe (MEPS)
Sigma–Aldrich (http://www.sigmaaldrich.com/)
Spark Holland (www.sparkholland.com/)
Thermo Fisher Scientific (www.thermofisher.com/)
Waters (www.waters.com/)

[a] As of April 2009. The products or material displayed are not endorsed by the author or the publisher of this present text.

is free, as is the ability to display tables of contents, bibliographic information and abstracts. However, 'full-text articles' are available in PDF and HTML formats but require a subscription fee for access – see Section 13.1.1 above.

- *Companies.* Suppliers of scientific equipment and extraction technique consumables provide 'on-line' catalogues and application notes which can be a useful source of information (see Table 13.2).

- *Institutions.* Most research organizations, professional bodies and universities have their own web pages. For example, The Royal Society of Chemistry in the UK (www.rsc.org) and the American Chemical Society (www.acs.org) have links to various sites of interest to chemists. Some other relevant web sites are given in Table 13.3.

- *Databases.* Sites such as the 'ISI Web of Knowledge' provide access to scientific publications: use these to find relevant literature for specific topics. Access is via the Web sites at http://wok.mimas.ac.uk/although you will need a username and password – check with your Department, School or library.

Table 13.3 Selected useful web sites[a]

Organization	Web address
American Chemical Society	http://www.acs.org
International Union of Pure and Applied Chemistry (IUPAC)	http://www.iupac.org/
Laboratory of the Government Chemist (LGC)	http://www.lgc.co.uk
National Institute of Standards and Technology (NIST) Laboratory	http://www.nist.gov
National Institute of Standards and Technology (NIST) WebBook	http://webbook.nist.gov
The Royal Society of Chemistry (RSC)	http://www.rsc.org
United States Environmental Protection Agency	http://www.epa.gov

[a] As of April 2009. The products or material displayed are not endorsed by the author or the publisher of this present text.

Summary

This final chapter highlights the different resources that are available to enable the reader to keep up-to-date with their studies/research. The developing role of the Worldwide Web in assisting this process is highlighted.

Responses to Self-Assessment Questions

Chapter 1

Response 1.1

A range of properties can be important when assessing organic compounds, including melting point, boiling point, molecular weight, dielectric constant and the octanol–water partition coefficient (K_{ow} or log P).

Response 1.2

Coning and quartering involves making a pile of the soil in a dome shape; making a cross on the top of the soil dome with a piece of sheet aluminium and removing the soil from opposite quarters of the cross. With these two new soil sub-samples, make a further soil dome shape (now obviously smaller in height than before) and repeat the process of quartering. This process is repeated until an appropriate sample size is obtained for the extraction step.

Response 1.3

Plastic containers are not recommended for aqueous samples as plasticizers are prone to leach from the vessels which can cause problems at later stages of the analysis, e.g. phthalates which are detected by gas chromatography (see Section 1.5.1).

Extraction Techniques in Analytical Sciences   John R. Dean
© 2009 John Wiley & Sons, Ltd

Response 1.4

In the TIC mode a mass spectrum of each eluting compound as well as a signal response is recorded. The derivation of a mass spectrum allows compound identification to take place via a dedicated PC-based database. In the SIM mode only selected ions representative of the compounds under investigation are monitored, leading to enhanced signal sensitivity.

Response 1.5

It may be possible to observe significant peak tailing (the peak appears to 'drag' out producing a non-Gaussian shaped peak) indicating the possibility of poor separation due to unreacted silanol groups.

Response 1.6

Calibration graphs are normally used to describe a relationship between two variables, x and y. It is normal practice to identify the x-axis as the horizontal axis (abscissa axis) and to use this for the independent variable, e.g. concentration (with its appropriate units). The vertical or ordinate axis (y-axis) is used to plot the dependent variable, e.g. signal response (with units, if appropriate). The mathematical relationship most commonly used for straight-line graphs is:

$$y = mx + c$$

where y is the signal response, e.g. signal (mV), x is the concentration of the working solution (in appropriate units, e.g. $\mu g \ ml^{-1}$ or ppm), m is the slope of the graph and c is the intercept on the x-axis.

A typical graphical representation of the data obtained from an experiment to determine the level of chlorobezene in a sample using chromatography is shown in Figure SAQ 1.6 (from the data tabulated in Table 1.3).

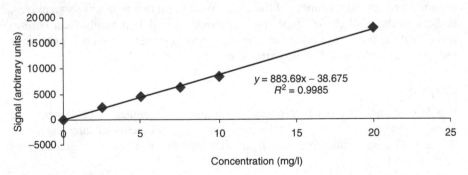

Figure SAQ 1.6 Calibration graph for chlorobenzene (cf. SAQ 1.6).

Response 1.7

Based on the equation for a straight line, $y = mx + c$, it was possible to calculate, using 'Excel', the values for the equation in SAQ 1.6, namely:

$$y = 883.69x - 38.675$$

Therefore, this equation can be re-arranged as follows to produce the concentration (x) of chlorobenzene in the original sample:

$$x = (1234 + 38.675)/883.69$$
$$= 1.4 \, mg/l$$

Response 1.8

The evaporation process may be increased by altering the:

- flow rate of the impinging gas (too high a rate and losses may occur);

- position of the impinger gas with respect to the extract surface;

- solvent extract surface area available for evaporation.

Chapter 2

Response 2.1

The answer is 4.

Response 2.2

In between each inversion, and while the stopper is in the palm of the hand, the stopcock is opened to release any gases that may build-up with the funnel. (Remember to close the stopcock before inverting the funnel again!)

Chapter 3

Response 3.1

In end-capping a further reaction is carried out on the residual silanols using a short-chain alkyl group to remove the hydroxyl groups. It is typical that the addition of a C1 moiety is indicative of end-capping (note: end-capping is not totally effective).

Response 3.2

A variation on this type of cartridge system or syringe filter is when a plunger is inserted into the cartridge barrel. In this situation the solvent is added to the syringe barrel and forced through the SPE system using the plunger. This system is effective if only a few samples are to be processed; for early method development, the SPE method is simple or useful when no vacuum system is available.

Response 3.3

The SPE disc, with its thin sorbent bed and large surface area, allows rapid flow rates of solvent. Typically, one litre of aqueous sample can be passed through an 'Empore' disc in approximately 10 min whereas with a cartridge system the same volume of aqueous sample may take approximately 100 min! However, large flow rates can result in poor recovery of the compound of interest due to there being a shorter time for compound–sorbent interaction.

Response 3.4

The general methodology for SPE is as follows.

Sorbent: C18

Wetting the sorbent: Pass 1.0 ml of methanol or acetonitrile per 100 mg of sorbent. This solvent has several functions, e.g. it will remove impurities from the sorbent that may have been introduced in the manufacturing process. In addition, as reversed phase sorbents are hydrophobic, they need the organic solvent to solvate or wet their surfaces.

Conditioning: Pass 1 ml of water or buffer per 100 mg of sorbent. Do not allow the sorbent to dry out before applying the sample.

Loading: A known volume of sample is loaded in a high polarity solvent or buffer. The solvent may be one that has been used to extract the compound from a solid matrix.

Rinsing: Unwanted, extraneous material is removed by washing the sample-containing sorbent with a high-polarity solvent or buffer. This process may be repeated.

Elution: Elute compounds of interest with a less polar solvent, e.g. methanol or the HPLC mobile phase (if this is the method of subsequent analysis); 0.5–1.0 ml per 100 mg of sorbent is typically required for elution.

Finally, the SPE cartridge or disc is discarded.

Response 3.5

The use of on-line SPE offers several advantages to the laboratory. For example, the number of manual manipulations decreases which improves the precision of the data, there is a lower risk of contamination as the system is closed from the point of sample injection through to the chromatographic output to waste, all of the compound loaded onto the pre-column is transferred to the analytical column and the analyst is available to perform other tasks.

Response 3.6

Once you find some key references to developments in the field of solid phase extraction in analytical sciences it might be worth considering how you might apply then to your studies/research in recovering organic compounds from a variety of matrices.

Chapter 4

Response 4.1

The SPME holder provides two functions, one is to provide protection for the fibre during transport while the second function is to allow piercing of the rubber septum of the gas chromatograph injector via a needle.

Response 4.2

In the case of HPLC, the fibre is inserted in a chamber that allows the mobile phase to affect desorption.

Response 4.3

Once you find some key references to developments in the field of solid phase microextraction in analytical sciences it might be worth considering how you might apply then to your studies/research in recovering organic compounds from a variety of matrices.

Chapter 5

Response 5.1

If the gas chromatograph is fitted with a PTV injector (see Section 1.5.1) then up to 50 µl of organic solvent can be used for microextraction.

Response 5.2

The needle with a suspended drop of organic solvent would be positioned in the headspace above an aqueous sample.

Response 5.3

Once you find some key references to developments in the field of membrane extraction in analytical sciences it might be worth considering how you might apply then to your studies/research in recovering organic compounds from a variety of matrices.

Chapter 6

Response 6.1

The initial process (Stage 1) (Figure SAQ 6.1) is slow, with respect to time, but leads to significant recovery of organic compounds from the sample matrix, due to three processes: desorption of organic compounds from matrix active sites; solvation of organic compounds by the (organic) solvent; diffusion of organic compounds through a static solvent layer. In contrast, Stage 2 (Figure SAQ 6.1) is (relatively) fast. In this stage, the organic compounds are rapidly removed from their initial matrix site by the flowing (bulk) solvent.

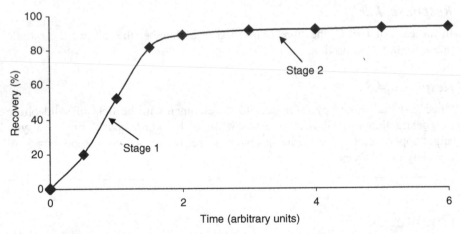

Figure SAQ 6.1 Typical extraction profile for the recovery of an organic compound from a solid matrix (cf. SAQ 6.1).

Response 6.2

In the case of the former, a localized effect is evident from the probe, whereas in the latter a more disperse effect is observed. In addition, the probe comes into contact with the sample and solvent, whereas in the case of the bath no such contact occurs.

Response 6.3

The actions of the various mechanical shakers available can be as follows:

- An *orbital* shaker – allows the sample/solvent to 'fall over itself' by the rotating action of the shaker.

- A *horizontal* shaker – allows the sample/solvent to interact primarily at the point of contact by the forward/back action of the shaker.

- A *rocking* shaker – allows the sample/solvent to interact at the point of contact by the twisting action of the shaker.

Response 6.4

Once you find some key references to developments in the field of ultrasonic extraction in analytical sciences it might be worth considering how you might apply then to your studies/research in recovering organic compounds from a variety of matrices.

Chapter 7

Response 7.1

A POP is an organic compound that survives in its original chemical form (or produces a significant breakdown product) in the environment for a considerable amount of time. Perhaps the most notorious and 'infamous' POP in this respect is DDT (and its metabolites, DDE and DDD) (see Figure 7.9 for the molecular structures of DDT and DDE).

Response 7.2

Well it obviously does as the scientific literature contains many examples of research scientists who have considered the PFE operating parameters.

Response 7.3

Once you find some key references to developments in the field of pressurized fluid extraction (pressurized liquid extraction or accelerated solvent extraction) in analytical sciences it might be worth considering how you might apply them to your studies/research in recovering organic compounds from a variety of matrices.

Chapter 8

Response 8.1

The heating effect in microwave cavities is due to the displacement of opposite charges, i.e. dielectric polarization; the most important one for microwaves is dipolar polarization. The polarization is achieved by the reorientation of permanent dipoles of compounds by the applied electric field. A polarized compound will rotate to align itself within the electric field at a rate of 2.45×10^9 s^{-1}.

Response 8.2

An explanation to this question can be proposed by considering the different heating methods being used between microwave and conventional heating methods. Figure 8.2 shows the typical heating mechanism when using a conventional approach, i.e. external heat, supplied by, for example, an isomantle to the external surface of the round-bottomed flask which causes conductive heating to take place. This results in convection currents being established within the solvent where warm solvent flows away from the internal edge of the flask to cooler regions until all of the solvent eventually gets warm/hot. In contrast, in a microwave heated approach (Figure 8.3) the process is very different: localized superheating occurs within the solvent within the flask resulting in no surface effects. As a result the organic solvent is heated much faster up to its boiling point. A direct comparison of conventional and microwave heating of distilled water is shown in Figure 8.4. It can be seen that the microwave-heated water quickly reaches the boiling point of water (approximately 6–7 min) whereas conventionally heated water takes much longer (approximately 15 min).

Response 8.3

Once you find some key references to developments in the field of microwave-assisted extraction in analytical sciences it might be worth considering how you might apply then to your studies/research in recovering organic compounds from a variety of matrices.

Chapter 9

Response 9.1

As the sorbent in a reversed phase solid phase extraction cartridge.

Response 9.2

As the sorbent in a reversed phase high performance liquid chromatography column.

Response 9.3

It will add a C1 moiety to the unreacted silanol groups on the surface of the silica.

Response 9.4

Once you find some key references to developments in the field of matrix solid phase dispersion in analytical sciences it might be worth considering how you might apply then to your studies/research in recovering organic compounds from a variety of matrices.

Chapter 10

Response 10.1

A phase diagram identifies regions where the substance occurs, as a result of temperature or pressure, as a single phase, i.e. a solid, liquid or gas. The divisions between these regions are bounded by curves indicating the co-existence of two phases.

Response 10.2

Once you find some key references to developments in the field of supercritical fluid extraction in analytical sciences it might be worth considering how you might apply then to your studies/research in recovering organic compounds from a variety of matrices.

Chapter 11

Response 11.1

Specific meteorological conditions include wind, rain, snow, draught, etc.

Response 11.2

'Tenax' is a common weak adsorbent composed of poly(2,6-diphenyl-*p*-phenylene oxide).

Response 11.3

Once you find some key references to developments in the field of air sampling in analytical sciences it might be worth considering how you might apply then to your studies/research in recovering organic compounds from a variety of matrices.

Chapter 12

Response 12.1

Sample mass This is often a balance between obtaining a representative and homogenous sample that can be extracted versus the total amount of sample available. In some cases, the amount of sample available may be large whereas in other cases only a limited quantity is available. If sample size is not a limiting factor, most of the extraction techniques have the capacity to handle samples of up to 10 g.

Extraction time The ability to extract samples rapidly needs to be considered with the ability of the technique to perform the extraction simultaneously (or not). Extractions can be performed rapidly using shake-flask, sonication, MAE and PFE. However, each particular extraction technique needs to be considered alongside other parameters. The ability of MAE to perform multiple sample extractions (up to 40) simultaneously offers the maximum benefit in this case.

Solvent type and consumption Most extraction techniques require organic solvents that are generally polar and contain chlorine, to solvate and recover organic compounds from sample matrices. In addition, with the exception of Soxhlet extraction, most approaches generally use small quantities of organic solvents which make them cost effective and potentially more environmentally friendly. However, the most influential technique in this case is supercritical fluid extraction which uses no organic solvent for recovery of organic compounds from matrices, unless a modifier is required for polar compounds.

Extraction method The use of elevated temperature is the most common single approach to facilitate recovery of organic compounds from sample matrices. In some instances the use of elevated temperatures and pressures enhances recovery of organic compounds in a shorter extraction times, e.g. MAE, PFE and SFE.

Sequential or simultaneous Soxhlet, shake-flask, sonication and MSPD can extract more than one sample simultaneously simply by multiplying the amount of apparatus required without significant additional costs being incurred. The major extraction technique that can perform simultaneous extractions is MAE with modern instruments being capable of recovering organic compounds from up to 40 samples.

Method development time A difficult question to answer as it primarily depends on the skill of the operator. However, a simple rule of thumb might indicate that the more instrumentation associated with the extraction technique, the more method development time is required.

Operator skill As above (see 'Method development time') the more instrumental approaches, e.g. MAE, SFE and PFE, often require more operator skill because of the complexity of operation and the potential for instrument failure/breakdown.

Equipment cost The 'more instrumental extraction techniques' have a higher capital purchase cost. In addition, the possibility of instrument failure/breakdown can also add to the running costs of such instruments. All capital apparatus costs need to be considered alongside running costs which can accumulate quickly with the prices for organic solvent, filters, replacement extraction vessels, frits, thimbles, cartridges, etc.

Level of automation Any amount of automation can reduce imprecision in the extraction process compared to manual operations. In addition, the use of automation can lead to enhanced productivity in the laboratory, i.e. more samples extracted per hour/per day, provided that the apparatus is appropriately maintained and regularly serviced to pre-empt breakdown/failure.

USEPA Method The existence of specific and dedicated analytical extraction procedures for most techniques provides an opportunity for reduced method development time and transfer of procedures (and hence data) between laboratories.

Response 12.2

Identifying some key reviews is a good starting point to readily acquire background information on the techniques described. The information acquired can then be applied in research projects, essay writing and other report preparation (being careful to avoid plagiarism).

Glossary of Terms

This section contains a glossary of terms, all of which are used in the text. It is not intended to be exhaustive, but to explain briefly those terms which often cause difficulties or may be confusing to the inexperienced reader.

Accelerated solvent extraction (ASE) Method of extracting analytes from matrices using solvent at elevated pressure and temperature (*see also* Pressurized fluid extraction).

Accuracy A quantity referring to the difference between the mean of a set of results or an individual result and the value which is accepted as the true or correct value for the quantity measured.

Analyte The component of a sample which is ultimately determined directly or indirectly.

Anion Ion having a negative charge; an atom with extra electrons. Atoms of non-metals, in solution, become anions.

Blowdown Removal of liquids and/or solids from a vessel by the use of pressure; often used to remove solvents to pre-concentrate the analyte.

BTEX Acronym used to describe the following volatile organic compounds: benzene, toluene, ethylbenzene and *ortho*-, *meta*- and *para*-xylenes.

Calibration The set of operations which establish, under specified conditions, the relationship between values indicated by a measuring instrument or measuring system and the corresponding known values of the measurand.

Calibration curve Graphical representation of measuring signal as a function of quantity of analyte.

Extraction Techniques in Analytical Sciences   John R. Dean
© 2009 John Wiley & Sons, Ltd

Cation Ion having a positive charge. Atoms of metals, in solution, become cations.

Certified Reference Material (CRM) Reference material, accompanied by a certificate, one or more of whose property values are certified by a procedure which establishes its traceability to an accurate realization of the unit in which the property values are expressed, and for which each certified value is accompanied by an uncertainty at a stated level of confidence.

Confidence interval Range of values that contains the true value at a given level of probability. The level of probability is called the confidence level.

Confidence limit The extreme values or end values in a confidence interval.

Contamination Contamination in trace analysis is the unintentional introduction of analyte(s) or other species which are not present in the original sample and which may cause an error in the determination. It can occur at any stage in the analysis. Quality assurance procedures such as analyses of blanks or of reference materials are used to check for contamination problems.

Control of Substances Hazardous to Health (COSHH) Regulations that impose specific legal requirements for risk assessment wherever hazardous chemicals or biological agents are used.

Dilution factor The mathematical factor applied to the determined value (data obtained from a calibration graph) that allows the concentration in the original sample to be determined. Frequently, for solid samples this will involve a sample weight and a volume to which the digested/extracted sample is made up to prior to analysis. For liquid samples this will involve an initial sample volume and a volume to which the digested/extracted sample is made up to prior to analysis.

Eluent The mobile liquid phase in liquid or in solid phase extraction.

Error The error of an analytical result is the difference between the result and a 'true' value.

>   *Random error* Result of a measurement minus the mean that would result from an infinite number of measurements of the same measurand carried out under repeatability conditions.

>   *Systematic error* Mean that would result from an infinite number of measurements of the same measurand carried out under repeatability conditions minus the true value of the measurand.

Extraction The removal of a soluble material from a solid mixture by means of a solvent or the removal of one or more components from a liquid mixture by use of a solvent with which the liquid is immiscible or nearly so.

Figure of merit A parameter that describes the quality of performance of an instrument or an analytical procedure.

Heterogeneity The degree to which a property or a constituent is randomly distributed throughout a quantity of material. The degree of heterogeneity is the determining factor of sampling error.

Homogeneity The degree to which a property or a constituent is uniformly distributed throughout a quantity of material. A material may be homogenous with respect to one analyte but heterogeneous with respect to another.

Interferent Any component of the sample affecting the final measurement.

Kuderna–Danish evaporator Apparatus for sample concentration consisting of a small (10 ml) graduated test tube connected directly beneath a 250 or 500 ml flask. A steam bath provides heat for evaporation with the concentrate collecting in the test tube.

Limit of detection The detection limit of an individual analytical procedure is the lowest amount of an analyte in a sample which can be detected but not necessarily quantified as an exact value. The limit of detection, expressed as the concentration c_L or the quantity q_L, is derived from the smallest measure, x_L that can be detected with reasonable certainty for a given procedure. The value x_L is given by the equation:

$$x_L = x_{bl} + k s_{bl}$$

where x_{bl} is the mean of the blank measures, s_{bl} is the standard deviation of the blank measures and k is a numerical factor chosen according to the confidence level required. For many purposes the limit of detection is taken to be $3 s_{bl}$ or $3 \times$ 'the signal-to-noise ratio', assuming a zero blank.

Limit of quantitation The limit of quantitation of an individual analytical procedure is the lowest amount of an analyte in a sample which can be quantitatively determined with suitable uncertainty. It may also be referred to as the limit of determination. The limit of quantitation can be taken as $10 \times$ 'the signal-to-noise ratio', assuming a zero blank.

Linear dynamic range (LDR) The concentration range over which the analytical working calibration curve remains linear.

Linearity Defines the ability of the method to obtain test results proportional to the concentration of analyte.

Liquid–liquid extraction A method of extracting a desired component from a liquid mixture by bringing the solution into contact with a second liquid, the solvent, in which the component is also soluble and which is immiscible with the first liquid or nearly so.

Matrix The carrier of the test component (analyte); all the constituents of the material except the analyte or the material with as low a concentration of the analyte as it is possible to obtain.

Measurand Particular quantity subject to measurement.

Method The overall, systematic procedure required to undertake an analysis. It includes all stages of the analysis, not just the (instrumental) end determination.

Microwave-assisted extraction (MAE) Method of extracting analytes from matrices using a solvent at elevated temperatures (and pressures) based on microwave radiation. Can be carried out in open or sealed vessels.

Microwave digestion Method of digesting an organic matrix to liberate metal content using an acid at elevated temperatures (and pressures) based on microwave radiation. Can be carried out in open or sealed vessels.

Outlier An outlier may be defined as an observation in a set of data that appears to be inconsistent with the remainder of that set.

Pesticide A pesticide is any substance or mixture of substances intended for preventing, destroying, repelling or mitigating any pest. Pests can be insects, mice and other animals, unwanted plants (weeds), fungi, or microorganisms like bacteria and viruses. Though often misunderstood to refer only to *insecticides*, the term pesticide also applies to herbicides, fungicides and various other substances used to control pests.

Polycyclic aromatic hydrocarbons (PAHs) These are a large group of organic compounds, comprising two or more aromatic rings, which are widely distributed in the environment.

Precision The closeness of agreement between independent test results obtained under stipulated conditions.

Pressurized fluid extraction (PFE) Method of extracting analytes from matrices using solvent at elevated pressures and temperatures (*see also* Accelerated solvent extraction).

Qualitative Qualitative analysis is chemical analysis designed to identify the components of a substance or mixture.

Quality assurance All those planned and systematic actions necessary to provide adequate confidence that a product or services will satisfy given requirements for quality.

Quality control The operational techniques and activities that are used to fulfil requirements of quality.

Quality control chart A graphical record of the monitoring of control samples which helps to determine the reliability of the results.

Quantitative Quantitative analysis is normally taken to mean the numerical measurement of one or more analytes to the required level of confidence.

Reagent A test substance that is added to a system in order to bring about a reaction or to see whether a reaction occurs (e.g. an analytical reagent).

Reagent blank A reagent blank is a solution obtained by carrying out all steps of the analytical procedure in the absence of a sample.

Recovery The fraction of the total quantity of a substance recoverable following a chemical procedure.

Reference material A material or substance, one or more of whose property values are sufficiently homogeneous and well established to be used for the calibration of an apparatus, the assessment of a measurement method, or for assigning values to materials.

Repeatability Precision under repeatability conditions, i.e. conditions where independent test results are obtained with the same method on identical test items in the same laboratory, by the same operator using the same equipment within short intervals of time.

Reproducibility Precision under reproducibility conditions, i.e. conditions where test results are obtained with the same method on identical test items in different laboratories with different operators using different equipment.

Robustness The robustness of an analytical procedure is a measure of its capacity to remain unaffected by small, but deliberate variations in method parameters and provides an indication of its reliability during normal usage. Sometimes it is referred to as *ruggedness*.

Rotary evaporation Removal of solvents by distillation under vacuum.

Sample A portion of material selected from a larger quantity of material. The term needs to be qualified, e.g. representative sample, sub-sample, etc.

Selectivity (in analysis) (i) Qualitative – the extent to which other substances interfere with the determination of a substance according to a given procedure. (ii) Quantitative – a term used in conjunction with another substantive (e.g. constant, coefficiemt, index, factor, number) for the quantitative characterization of interferences.

Sensitivity The change in the response of a measuring instrument divided by the corresponding change in stimulus.

Shake-flask extraction Method of extracting analytes from matrices using agitation or shaking in the presence of a solvent.

Signal-to-noise ratio A measure of the relative influence of noise on a control signal. Usually taken as the magnitude of the signal divided by the standard deviation of the background signal.

Solid-phase extraction (SPE) A sample preparation technique that uses a solid-phase packing contained in a small plastic cartridge. The solid stationary phases are the same as HPLC packings; however, the principle is different from HPLC. The process as most often practiced requires four steps: conditioning the sorbent, adding the sample, washing away the impurities and eluting the sample in as small a volume as possible with a strong solvent.

Solid-phase microextraction (SPME) A sample preparation technique that uses a fused silica fibre coated with a polymeric phase to sample either an aqueous solution or the headspace above a sample. Analytes are absorbed by the polymer coating and the SPME fibre is directly transferred to a GC injector or special HPLC injector for desorption and analysis.

Solvent extraction The removal of a soluble material from a solid mixture by means of a solvent or the removal of one or more components from a liquid mixture by use of a solvent with which the liquid is immiscible or nearly so.

Soxhlet extraction Equipment for the continuous extraction of a solid by a solvent. The material to be extracted is placed in a porous cellulose thimble, and continually condensing solvent is allowed to percolate through it, and return to the boiling vessel, either continuously or intermittently.

Specificity The ability of a method to measure only what it is intended to measure. Specificity is the ability to assess unequivocally the analyte in the presence of components which may be expected to be present. Typically these might include impurities, degradants, matrices, etc.

Spiked sample 'Spiking a sample' is a widely used term taken to mean the addition of a known quantity of analyte to a matrix which is close to or identical with that of the samples of interest.

Standard (general) A standard is an entity established by consensus and approved by a recognized body. It may refer to a material or solution (e.g. an organic compound of known purity or an aqueous solution of a metal of agreed concentration) or a document (e.g. a methodology for an analysis or a quality system). The relevant terms are:

> *Analytical standard* (*also known as* **Standard solution**) A solution or matrix containing the analyte which will be used to check the performance of the method/instrument.

> *Calibration standard* The solution or matrix containing the analyte (measurand) at a known value with which to establish a corresponding response from the method/instrument.

External standard A measurand, usually identical with the analyte, analysed separately from the sample.

Internal standard A measurand, similar to but not identical with the analyte is combined with the sample.

Standard method A procedure for carrying out a chemical analysis which has been documented and approved by a recognized body.

Standard addition The addition of a known amount of analyte to the sample in order to determine the relative response of the detector to an analyte within the sample matrix. The relative response is then used to assess the sample analyte concentration.

Stock solution A stock solution is generally a standard or reagent solution of known accepted stability, which has been prepared in relatively large amounts of which portions are used as required. Frequently such portions are used following further dilution.

Sub-sample A subsample may be (i) a portion of the sample obtained by selection or division, (ii) an individual unit of the lot taken as part of the sample or (iii) the final unit of multistage sampling.

Supercritical fluid extraction (SFE) Method of extracting analytes from matrices using a supercritical fluid at elevated pressures and temperatures. A supercritical fluid is any substance above its critical temperature and critical pressure.

True value A value consistent with the definition of a given particular quantity

Ultrasonic extraction Method of extracting analytes from matrices with a solvent using either an ultrasonic bath or probe

Uncertainty Parameter associated with the result of a measurement that characterizes the dispersion of the values that could reasonably be attributed to the measurand.

SI Units and Physical Constants

SI Units

The SI system of units is generally used throughout this book. It should be noted, however, that according to present practice, there are some exceptions to this, for example, wavenumber (cm^{-1}) and ionization energy (eV).

Base SI units and physical quantities

Quantity	Symbol	SI Unit	Symbol
length	l	metre	m
mass	m	kilogram	kg
time	t	second	s
electric current	I	ampere	A
thermodynamic temperature	T	kelvin	K
amount of substance	n	mole	mol
luminous intensity	I_v	candela	cd

Prefixes used for SI units

Factor	Prefix	Symbol
10^{21}	zetta	Z
10^{18}	exa	E
10^{15}	peta	P

(continued overleaf)

Extraction Techniques in Analytical Sciences   John R. Dean
© 2009 John Wiley & Sons, Ltd

Prefixes used for SI units (*continued*)

Factor	Prefix	Symbol
10^{12}	tera	T
10^{9}	giga	G
10^{6}	mega	M
10^{3}	kilo	k
10^{2}	hecto	h
10	deca	da
10^{-1}	deci	d
10^{-2}	centi	c
10^{-3}	milli	m
10^{-6}	micro	μ
10^{-9}	nano	n
10^{-12}	pico	p
10^{-15}	femto	f
10^{-18}	atto	a
10^{-21}	zepto	z

Derived SI units with special names and symbols

Physical quantity	SI unit		Expression in terms of base or derived SI units
	Name	Symbol	
frequency	hertz	Hz	$1\,\text{Hz} = 1\,\text{s}^{-1}$
force	newton	N	$1\,\text{N} = 1\,\text{kg}\,\text{m}\,\text{s}^{-2}$
pressure; stress	pascal	Pa	$1\,\text{Pa} = 1\,\text{Nm}^{-2}$
energy; work; quantity of heat	joule	J	$1\,\text{J} = 1\,\text{Nm}$
power	watt	W	$1\,\text{W} = 1\,\text{J}\,\text{s}^{-1}$
electric charge; quantity of electricity	coulomb	C	$1\,\text{C} = 1\,\text{A}\,\text{s}$
electric potential; potential difference; electromotive force; tension	volt	V	$1\,\text{V} = 1\,\text{J}\,\text{C}^{-1}$
electric capacitance	farad	F	$1\,\text{F} = 1\,\text{C}\,\text{V}^{-1}$
electric resistance	ohm	Ω	$1\,\Omega = 1\,\text{V}\,\text{A}^{-1}$
electric conductance	siemens	S	$1\,\text{S} = 1\,\Omega^{-1}$
magnetic flux; flux of magnetic induction	Weber	Wb	$1\,\text{Wb} = 1\,\text{V}\,\text{s}$
magnetic flux density;	tesla	T	$1\,\text{T} = 1\,\text{Wb}\,\text{m}^{-2}$
magnetic induction inductance	henry	H	$1\,\text{H} = 1\,\text{Wb}\,\text{A}^{-1}$

(*continued overleaf*)

Derived SI units with special names and symbols (*continued*)

Physical quantity	SI unit		Expression in terms of
	Name	Symbol	base or derived SI units
Celsius temperature	degree Celsius	°C	$1\,°\mathrm{C} = 1\,\mathrm{K}$
luminous flux	lumen	lm	$1\,\mathrm{lm} = 1\,\mathrm{cd\,sr}$
illuminance	lux	lx	$1\,\mathrm{lx} = 1\,\mathrm{lm\,m^{-2}}$
activity (of a radionuclide)	becquerel	Bq	$1\,\mathrm{Bq} = 1\,\mathrm{s^{-1}}$
absorbed dose; specific energy	gray	Gy	$1\,\mathrm{Gy} = 1\,\mathrm{J\,kg^{-1}}$
dose equivalent	sievert	Sv	$1\,\mathrm{Sv} = 1\,\mathrm{J\,kg^{-1}}$
plane angle	radian	rad	1^a
solid angle	steradian	sr	1^a

[a] rad and sr may be included or omitted in expressions for the derived units.

Physical Constants

Recommended values of selected physical constants[a]

Constant	Symbol	Value
acceleration of free fall (acceleration due to gravity)	g_n	$9.806\ 65\ \mathrm{ms^{-2}}$[b]
atomic mass constant (unified atomic mass unit)	m_u	$1.660\ 540\ 2(10) \times 10^{-27}\ \mathrm{kg}$
Avogadro constant	L, N_A	$6.022\ 136\ 7(36) \times 10^{23}\ \mathrm{mol^{-1}}$
Boltzmann constant	k_B	$1.380\ 658(12) \times 10^{-23}\ \mathrm{J\,K^{-1}}$
electron specific charge (charge-to-mass ratio)	$-e/m_e$	$-1.758\ 819 \times 10^{11}\ \mathrm{Ckg^{-1}}$
electron charge (elementary charge)	e	$1.602\ 177\ 33(49) \times 10^{-19}\ \mathrm{C}$
Faraday constant	F	$9.648\ 530\ 9(29) \times 10^4\ \mathrm{C\,mol^{-1}}$
ice-point temperature	T_{ice}	$273.15\ \mathrm{K}$[b]
molar gas constant	R	$8.314\ 510(70)\ \mathrm{JK^{-1}\,mol^{-1}}$
molar volume of ideal gas (at 273.15 K and 101 325 Pa)	V_m	$22.414\ 10(19) \times 10^{-3}\ \mathrm{m^3\,mol^{-1}}$
Planck constant	h	$6.626\ 075\ 5(40) \times 10^{-34}\ \mathrm{J\,s}$
standard atmosphere	atm	$101\ 325\ \mathrm{Pa}$[b]
speed of light in vacuum	c	$2.997\ 924\ 58 \times 10^8\ \mathrm{ms^{-1}}$[b]

[a] Data are presented in their full precision, although often no more than the first four or five significant digits are used; figures in parentheses represent the standard deviation uncertainty in the least significant digits.
[b] Exactly defined values.

The Periodic Table

Legend:

3	— Atomic number
0.98	— Pauling electronegativity
Li	— Element
6.941	— Atomic weight (^{12}C)

Group 1	Group 2	Group 3	4	5	6	7	8	9	10	11	12	Group 13	Group 14	Group 15	Group 16	Group 17	Group 18
1 2.20 **H** 1.008																	2 **He** 4.003
3 0.98 **Li** 6.941	4 1.57 **Be** 9.012											5 2.04 **B** 10.811	6 2.55 **C** 12.011	7 3.04 **N** 14.007	8 3.44 **O** 15.999	9 3.98 **F** 18.998	10 **Ne** 20.179
11 0.93 **Na** 22.990	12 1.31 **Mg** 24.305											13 1.61 **Al** 26.98	14 1.90 **Si** 28.086	15 2.19 **P** 30.974	16 2.58 **S** 32.064	17 3.16 **Cl** 35.453	18 **Ar** 39.948
19 0.82 **K** 39.102	20 1.00 **Ca** 40.08	21 **Sc** 44.956	22 **Ti** 47.90	23 **V** 50.941	24 **Cr** 51.996	25 **Mn** 54.938	26 **Fe** 55.847	27 **Co** 58.933	28 **Ni** 58.71	29 **Cu** 63.546	30 **Zn** 65.37	31 1.81 **Ga** 69.72	32 2.01 **Ge** 72.59	33 2.18 **As** 74.922	34 2.55 **Se** 78.96	35 2.96 **Br** 79.909	36 **Kr** 83.80
37 0.82 **Rb** 85.47	38 0.95 **Sr** 87.62	39 **Y** 88.906	40 **Zr** 91.22	41 **Nb** 92.906	42 **Mo** 95.94	43 **Tc** (99)	44 **Ru** 101.07	45 **Rh** 102.91	46 **Pd** 106.4	47 **Ag** 107.87	48 **Cd** 112.40	49 1.78 **In** 114.82	50 1.96 **Sn** 118.69	51 2.05 **Sb** 121.75	52 2.10 **Te** 127.60	53 2.66 **I** 126.90	54 **Xe** 131.30
55 0.79 **Cs** 132.91	56 0.89 **Ba** 137.34	57 **La** 138.91	72 **Hf** 178.49	73 **Ta** 180.95	74 **W** 183.85	75 **Re** 186.2	76 **Os** 190.2	77 **Ir** 192.22	78 **Pt** 195.09	79 **Au** 196.97	80 **Hg** 200.59	81 2.04 **Tl** 204.37	82 2.32 **Pb** 207.19	83 2.02 **Bi** 208.98	84 **Po** (210)	85 **At** (210)	86 **Rn** (222)
87 **Fr** (223)	88 **Ra** 226.025	89 **Ac** 227.0	104 **Rf** (261)	105 **Db** (262)	106 **Sg** (263)	107 **Bh** (262)	108 **Hs** (263)	109 **Mt**	110 **Uun**	111 **Uuu**	112 **Unb**						

d transition elements

Lanthanides:

58 **Ce** 140.12	59 **Pr** 140.91	60 **Nd** 144.24	61 **Pm** (147)	62 **Sm** 150.35	63 **Eu** 151.96	64 **Gd** 157.25	65 **Tb** 158.92	66 **Dy** 162.50	67 **Ho** 164.93	68 **Er** 167.26	69 **Tm** 168.93	70 **Yb** 173.04	71 **Lu** 174.97

Actinides:

90 **Th** 232.04	91 **Pa** (231)	92 **U** 238.03	93 **Np** (237)	94 **Pu** (242)	95 **Am** (243)	96 **Cm** (247)	97 **Bk** (247)	98 **Cf** (249)	99 **Es** (254)	100 **Fm** (253)	101 **Md** (253)	102 **No** (256)	103 **Lw** (260)

General Index

Application Index

A wide range of applications are covered in this book, ranging from brief summaries in Chapters 6, 8, 9 and 10 (specifically Tables 6.1, 8.3, 9.1 and 10.2) through to more detailed explanations and data as detailed below.

Pressurized fluid extraction (PFE)

- Organochlorine pesticides from soil, 157
- PCBs, PCDDs and PCDFs from fish oil, 158
- Pharmaceuticals from sewage sludge, 154
- p,p'-DDT and p,p'-DDE from aged soils, 152
- Sulfamide antibiotics from aged agricultural soils, 155

Solid phase extraction (SPE)

Automated on-line:

- Sulphonamide antibiotics, neutral and acidic pesticides in natural waters, 78

Ion exchange:

- Alkylphenols from produced water from offshore oil installations, 66

Extraction Techniques in Analytical Sciences   John R. Dean
© 2009 John Wiley & Sons, Ltd

Solid phase microextraction (SPME)